ANIMAL TRACKS
of the
APPALACHIANS

Tamara Eder

with contributions from Ian Sheldon

LONE PINE

© 2001 by Lone Pine Publishing
First printed in 2001   10 9 8 7 6 5 4 3 2 1
Printed in Canada

THE PUBLISHER: LONE PINE PUBLISHING

1901 Raymond Avenue SW, Suite C	10145-81 Avenue
Renton, WA 98055	Edmonton, AB T6E 1W9
USA	Canada

Website: http://www.lonepinepublishing.com

National Library of Canadian Cataloguing in Publication Data

Eder, Tamara, (date)
   Animal tracks of the Appalachians

   Includes bibliographical references and index.
   ISBN 1-55105-256-3

   1. Animal tracks—Appalachian Region—Identification. I. Sheldon, Ian, (date)
II. Title.
QL768.E37 2000   591.47'9   C00-910473-9

Editorial Director: Nancy Foulds
Editor: Volker Bodegom
Proofreaders: Eli MacLaren, Lee Craig
Production Manager: Jody Reekie
Design, layout and production: Volker Bodegom, Monica Triska
Cover Design: Robert Weidemann
Technical Contributor: Mark Elbroch
Cartography: Volker Bodegom
Animal illustrations: by Gary Ross, except for one by Kindrie Grove (p. 99)
Track illustrations: Ian Sheldon
Cover illustration: White-tailed Deer by Gary Ross
Scanning: Elite Lithographers Co.

We acknowledge the financial support of the Government of Canada
through the Book Publishing Industry Development Program (BPIDP)
for our publishing activities.

PC: P4

CONTENTS

INTRODUCTION

If you have ever spent time with an experienced tracker, or perhaps a veteran hunter, then you know just how much there is to learn about the subject of tracking and just how exciting the challenge of tracking animals can be. Maybe you think that tracking is no fun, because all you get to see is the animal's prints. What about the animal itself—is that not much more exciting? Well, for most of us who don't spend a great deal of time in the beautiful wilderness of the Appalachians, the chances of seeing the elusive Bobcat or the fun-loving River Otter are slim. The closest that we may ever get to some animals will be through their tracks, and they can inspire a very intimate experience. Remember, you are following in the footsteps of the unseen—animals that are in pursuit of prey, or perhaps being pursued as prey.

This book offers an introduction to the complex world of tracking animals. Sometimes tracking is easy. At other times it is an incredible challenge that leaves you wondering just what animal made those unusual tracks. Take this book into the field with you, and it can provide some help with the first steps to identification. Animals tracks and trails are this book's focus; you will learn to recognize subtle differences for both. There are, of course, many additional signs to consider, such as scat and food caches, all of which help you to understand the animal that you are tracking.

Remember, it takes many years to become an expert tracker. Tracking is one of those skills that grows with you as you acquire new knowledge in new situations. Most importantly, you will have an intimate experience with nature. You will learn the secrets of the seldom seen. The more you discover, the more you will want to know, and by developing a good understanding of tracking, you will gain an excellent appreciation of the intricacies and delights of our marvelous natural world.

How to Use This Book

Most importantly, take this book into the field with you! Relying on your memory is not an adequate way to identify tracks. Track identification has to be done in the field, or with detailed sketches and notes that you can take home. Much of the process of identification involves circumstantial evidence, so you will have much more success when standing beside the track.

Feral Pig

Eastern Tiger Salamander

This book is laid out in an easy-to-use format. There is a quick reference appendix to the tracks of all the animals illustrated in the book (beginning on p. 140). This appendix is a fast way to familiarize yourself with certain tracks, and it guides you to the more informative descriptions of each animal and its tracks.

Each animal's description is illustrated with the appropriate footprints and the track patterns that it usually leaves. Although these illustrations are not exhaustive, they do show the tracks or groups of prints that you will most likely see. You will find a list of dimensions for the tracks, giving the general range, but there will always be extremes, just as there are with people who have unusually small or large feet. Under the category 'Size' (of animal), the 'greater-than' sign (>) is used when the size difference between the sexes is pronounced.

If you think that you may have identified a track, check the 'Similar Species' section. This section is designed to help you confirm your conclusions by pointing out other animals that leave similar tracks and showing you ways to distinguish among them.

CANADA

ONTARIO

MICHIGAN

NEW YORK

Albany

PENNSYLVANIA

Harrisburg

INDIANA

OHIO

NEW
JERSEY

WEST
VIRGINIA

DELAWARE

Charleston

MARYLAND

Frankfort

Richmond

KENTUCKY

VIRGINIA

Nashville

Raleigh

TENNESSEE

NORTH CAROLINA

Columbia

Atlanta

SOUTH
CAROLINA

N

ALABAMA

0   200 km
0   200 mi

GEORGIA

FLORIDA

Atlantic   Ocean

7

As you read this book, you will notice an abundance of words such as 'often,' 'mostly' and 'usually.' Unfortunately, tracking will never be an exact science; we cannot expect animals to conform to our expectations, so be prepared for the unpredictable.

Tips on Tracking

As you flip through this guide, you will notice clear, well-formed prints. Do not be deceived! It is a rare track that will ever show so clearly. For a good, clear print, the perfect conditions are slightly wet, shallow snow that isn't melting, or slightly soft mud that isn't actually wet. These conditions can be rare—most often you will be dealing with incomplete or faint prints where you cannot even really be sure of the number of toes.

Should you find yourself looking at a clear print, then the job of identification is much easier. There are a number of key features to look for: measure the length and width of the print, count the number of toes, check for claw marks and note how far away they

Porcupine

are from the body of the print, and look for a heel. Keep in mind more subtle features, such as the spacing between the toes, whether or not they are parallel, and whether fur on the sole of the foot has made the print less clear.

When you are faced with the challenge of identifying an unclear print—or even if you think that you have made a successful identification from one print alone— look beyond the single footprint and search out others. Do not rely on the dimensions of one print alone, but collect measurements from several prints to get an average impression. Even the prints within one trail can show a lot of variation.

Try to determine which is the fore print and which is the hind, and remember that many animals are built very differently from humans, having larger forefeet than hind feet. Sometimes the prints will overlap, or they can be directly on top of one another in a direct register. For some animals, the fore and hind prints are pretty much the same.

Check out the pattern that the tracks make together in the trail and follow the trail for as many paces as is necessary for you to become familiar with the pattern. Patterns are very important and can be the distinguishing feature between different animals with otherwise similar tracks. Follow the trail for some distance—it may give you some vital clues. For example, the trail may lead you to a tree, indicating that the animal is a climber, or it may lead down into a burrow. This part of tracking can be the most

Masked Shrew

Common Raven

rewarding, because you are following the life of the animal as it hunts, runs, walks, jumps, feeds or tries to escape a predator.

Take into consideration the habitat. Sometimes very similar species can be distinguished by their habitats only—one might be found on the riverbank, whereas another might be encountered just in the dense forest.

Think about your geographical location, too, because some animals have a limited range. This consideration can rule out some species and help you with your identification.

Remember that every animal will at some point leave a print or trail that looks just like the print or trail of a completely different animal!

Finally, keep in mind that if you track quietly, you might catch up with the maker of the prints.

 11

Terms & Measurements

Some of the terms used in tracking can be rather confusing, and they often depend on personal interpretation. For example, what comes to your mind if you see the word 'hopping'? Perhaps you see a person hopping about on one leg, or perhaps you see a rabbit hopping through the countryside. Clearly, one person's perception of motion can be very different from another's. Some useful terms are explained below, to clarify what is meant in this book and, where appropriate, how the measurements given fit in with each term.

The following terms are sometimes used loosely and interchangeably—for example, a rabbit might be described as 'a hopper' and a squirrel as 'a bounder,' yet both leave the same pattern of prints in the same sequence.

Ambling: Fast, rolling walking.

Bounding: A gait of four-legged animals in which the two hind feet land simultaneously, usually registering in front of the forefeet. Common in rodents and the rabbit family. 'Hopping' or 'jumping' can often be substituted.

hind prints

fore prints

Gait: An animal's gait describes how it is moving at some point in time and results in observable trail characteristics.

Galloping: A gait used by animals with four legs of even length, such as dogs, moving at high speed, hind feet registering in front of forefeet.

hind prints   fore prints

gallop group

Hopping: Similar to bounding. With four-legged animals it is usually indicated by tight clusters of prints, fore prints set between and behind the hind prints. A bird hopping on two feet creates a series of paired tracks along its trail.

Loping: Like galloping, but slower, with each foot falling independently and leaving a trail pattern that consists of groups of tracks in the sequence fore-hind-fore-hind, usually roughly in a line.

Mustelids (weasel family) often use ***2×2 loping***, in which the hind feet register directly on the fore prints. The resulting pattern has angled, paired tracks.

Running: Like galloping, but applied generally to animals moving at high speed. Also used for two-legged animals.

Trotting: Faster than walking, slower than running. The diagonally opposite limbs move simultaneously; that is, the right forefoot with the left hind, then the left forefoot with the right hind. This gait is the natural one for canids (dog family), short-tailed shrews and voles.

hind print

fore print

Canids may use ***side-trotting,*** a fast trotting in which the hind end of the animal shifts to one side. The resulting track pattern has paired tracks, with all the fore prints on one side and all the hind prints on the other.

Walking: A slow gait in which each foot moves independently of the others, resulting in an alternating track pattern. This gait is common for felines (cat family) and deer, as well as wide-bodied animals, such as bears and porcupines. The term is also used for two-legged animals.

Other Tracking Terms:

Dewclaws: Two small, toe-like structures set above and behind the main foot of most hoofed animals.

Direct Register: The hind foot falls directly on the fore print.

double register   *direct register*

Double Register: The hind foot overlaps the fore print only slightly or falls beside it, so that both prints can be seen at least in part.

Dragline: A line left in snow or mud by a foot or the tail dragging over the surface.

dragline

Gallop Group: A track pattern of four prints made at a gallop, usually with hind feet registering in front of forefeet.

Height: Taken at the animal's shoulder.

Length: The animal's body length from head to rump, not including the tail, unless otherwise indicated.

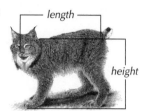

Metacarpal Pad: A small pad near the palm pad or between the palm pad and heel on the forefeet of bears and mustelids (members of the weasel family).

Print (also called '*track*'): Fore and hind prints are treated individually. Print dimensions given are 'length' (including claws—maximum values may represent occasional heel register for some animals) and 'width.' A group of prints made by each of the animal's feet makes up a track pattern.

Register: To leave a mark—said about a foot, claw or other part of an animal's body.

Retractable: Describes claws that can be pulled in to keep them sharp, as with felids (the cat family); these claws do not register in the prints. Foxes have semi-retractable claws.

Sitzmark: The mark left on the ground by an animal falling or jumping from a tree.

Straddle: The total width of the trail, with all prints considered.

Stride: For consistency among different animals, the stride is taken as the distance from the center of one print (or print group) to the center of the next one. Some books may use the term 'pace.'

Track: Same as 'print.'

Track Pattern: The pattern left after each foot registers once; a set of prints, such as a gallop group.

Trail: A series of track patterns; think of it as the path of the animal.

Southern Flying Squirrel

MAMMALS

River Otter

Moose

Fore and Hind Prints
Length: 4–7 in (10–18 cm)
   with dewclaws: to 11 in (28 cm)
Width: 3.5–6 in (9–15 cm)

Straddle
8.5–20 in (22–50 cm)

Stride
Walking: 1.5–3 ft (45–90 cm)
Trotting: to 4 ft (1.2 m)

Size (bull>cow)
Height: 5–6.5 ft (1.5–2 m)
Length: 7–8.5 ft (2.1–2.6 m)

Weight
600–1100 lb (270–500 kg)

walking

MOOSE
Alces alces

Once common in the northern Appalachians, the moose has dramatically declined because of habitat loss and hunting. If you find positive signs of the moose in the north, it's a discovery worth reporting to the nearest wildlife authority.

The ungainly shaped Moose moves gracefully, leaving a neat alternating walking pattern. The hind prints direct register or double register on the fore prints. Dewclaws— which give extra support for the animal's great weight— register in prints more than 1.2 inches (3 cm) deep, but far behind the hoof. In summer, look for tracks in mud beside ponds and other wet areas, where Moose especially like to feed; they are excellent swimmers. In winter, Moose feed in willow flats and coniferous forests, leaving a distinct browseline (highline). Ripped stems and scraped bark, 6 feet (1.8 m) or more above the ground, are additional signs of Moose.

Similar Species: White-tailed Deer (p. 22) tracks— which are smaller, with a narrower straddle and more foot drag—may be mistaken for a juvenile Moose's.

White-tailed Deer

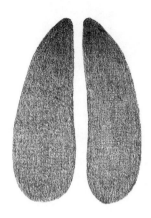

Fore and Hind Prints
Length: 2–3.5 in (5–9 cm)
Width: 1.6–2.5 in (4–6.5 cm)

Straddle
5–10 in (13–25 cm)

Stride
Walking: 10–20 in (25–50 cm)
Galloping: 6–15 ft (1.8–4.5 m)

Size (buck>doe)
Height: 3–3.5 ft (90–110 cm)
Length: to 6.3 ft (1.9 m)

Weight
120–350 lb (55–160 kg)

walking   *gallop group*

WHITE-TAILED DEER
Odocoileus virginianus

The keen hearing of this deer guarantees that it knows about you before you know about it. Frequently, all that we see is its conspicuous white tail in the distance as it gallops away, earning this deer the nickname 'flagtail.' This adaptable deer may be found throughout the region, in small groups at the edges of forests and in brushlands. The White-tailed Deer can be common around ranches and residential areas.

This deer's prints are heart-shaped and pointed. Its alternating walking track pattern shows the hind prints direct registered or double registered on the fore prints. In snow, or when a deer gallops on soft surfaces, the dewclaws register. This flighty deer gallops in the usual style, leaving hind prints ahead of fore prints, with toes spread wide for better footing.

Similar Species: Juvenile Moose (p. 20) tracks may be confused with large deer tracks. Feral Pig (p. 24) tracks are less pointed and have a shorter stride.

Feral Pig

**Fore and Hind Prints
(with dewclaws)**
Length: 2.5–3 in (6.5–7.5 cm)
Width: 2.3 in (5.8 cm)
Straddle
5–6 in (13–15 cm)
Stride
Trotting: 16–20 in (40–50 cm)
Size (male>female)
Height: 3 ft (90 cm)
Length: 4.3–6 ft (1.3–1.8 m)
Weight
77–440 lb (35–200 kg)

trotting

FERAL PIG (Wild Pig, Wild Boar)

Sus scrofa

Descended from introduced European animals, the Feral Pig has interbred with escaped domestic pigs. Populations of this sturdy beast of the dense undergrowth can be found scattered at low elevations in the central and southern Appalachians. These pigs, with their tusks, can be quite threatening.

A Feral Pig's print shows two prominent, widely spaced toe marks, and usually, except on firm surfaces, a clear, pointed dewclaw mark off to each side. The hind print is slightly smaller than the fore print. Feral Pigs are keen foragers, so their tracks can often be numerous, especially when they travel in a group. The Feral Pig usually trots—a typical alternating track pattern shows a double register of hind print on fore print. Other signs are wallows and diggings.

Similar Species: White-tailed Deer (p. 22) tracks are similar, but pointier, with a narrower gap between the toes, dewclaw marks to the rear (not the side) and a longer stride. The Domestic Pig (also *S. scrofa*) has a wider straddle and less neat tracks that often form two separate lines.

Horse

**Fore Print
(hind print is slightly smaller)**
Length: 4.5–6 in (11–15 cm)
Width: 4.5–5.5 in (11–14 cm)

Straddle
2–7.5 in (5–19 cm)

Stride
Walking: 17–28 in (43–70 cm)

Size
Height: to 6 ft (1.8 m)

Weight
to 1500 lb (680 kg)

walking

HORSE
Equus caballus

Widespread use of the popular Horse means that you can expect its tracks to show up almost anywhere.

Unlike any other animal in this book, the Horse has only one huge toe. This toe leaves an oval print with a distinctive 'frog' (V-shaped mark) at its base. When the Horse is shod, the horseshoe shows up clearly as a firm wall at the outside of the print. Not all horses are shod, so do not expect to see this outer wall on every horse print. A typical, unhurried horse trail shows an alternating walking pattern, with the hind prints registered on or behind the slightly larger fore prints. Horses are capable of a range of speeds—up to a full gallop—but most recreational horseback riders take a more leisurely outlook on life, preferring to walk their horses and soak up the mountain views!

Similar Species: Mules (rarely shod) have smaller tracks.

Black Bear

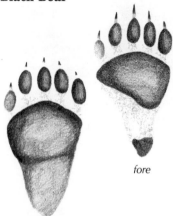

hind

fore

Fore Print
Length: 4–6.3 in (10–16 cm)
Width: 3.8–5.5 in (9.5–14 cm)

Hind Print
Length: 6–7 in (15–18 cm)
Width: 3.5–5.5 in (9–14 cm)

Straddle
9–15 in (23–38 cm)

Stride
Walking: 17–23 in (43–58 cm)

Size (male>female)
Height: 3–3.5 ft (90–110 cm)
Length: 5–6 ft (1.5–1.8 m)

Weight
200–600 lb (90–270 kg)

walking

BLACK BEAR
Ursus americanus

The Black Bear has a reduced and scattered distribution throughout the Appalachians' forested areas, but do not expect to see its tracks in winter, when it sleeps deeply. Finding fresh bear tracks can be a thrill, but take care—the bear may be just ahead. Never underestimate the potential power of a surprised bear!

Black Bear prints resemble small human prints, but they are wider and show claw marks. The small inner toe rarely registers. The forefoot's small heel pad often shows in the print, and the hind print shows a big heel. The bear's slow walk results in a slightly pigeon-toed double register with the hind print on the fore print. More frequently, at a faster pace, the hind foot oversteps the forefoot. When a bear runs, the two hind feet register in front of the forefeet in an extended cluster. Along well-worn bear paths, look for 'digs'—patches of dug-up earth—and 'bear trees' whose scratched bark shows that these bears climb.

Similar Species: The Black Bear is the only bear in this region.

29

Coyote

fore

hind

**Fore Print
(hind print is slightly smaller)**
Length: 2.4–3.2 in (6–8 cm)
Width: 1.6–2.4 in (4–6 cm)

Straddle
4–7 in (10–18 cm)

Stride
Walking: 8–16 in (20–40 cm)
Trotting: 17–23 in (43–58 cm)
Galloping/Leaping:
    2.5–10 ft (0.8 m–3 m)

Size (female is slightly smaller)
Height: 23–26 in (58–65 cm)
Length: 32–40 in (80–100 cm)

Weight
20–50 lb (9–23 kg)

*walking or
trotting*

*gallop
group*

COYOTE
(Brush Wolf, Prairie Wolf)
Canis latrans

This widespread, adaptable canine prefers open grasslands or woodlands. It hunts rodents and larger prey, either on its own, with a mate or in a family pack. If you find a coyote den—usually a wide-mouthed tunnel leading into a nesting chamber—do not bother the family or the female will have to move her pups to a safer location.

The oval fore prints are slightly larger than the hind prints. The fore heel pad is more triangular than the hind heel pad, which rarely registers clearly. The two outer toes usually do not register claw marks. The Coyote typically walks or trots in an alternating pattern; the walk has a wider straddle, and the trotting trail is often very straight. When it gallops, the Coyote's hind feet fall in front of its forefeet; the faster it goes, the straighter the gallop group. The Coyote's tail, which hangs down, leaves a dragline in deep snow.

Similar Species: A Domestic Dog's (*C. familiaris*) less oval prints splay more, and its trail is erratic. Foot hairs make Red Fox (p. 32) prints (usually smaller) less clear. Gray Fox (p. 34) tracks are much smaller.

Red Fox

fore

hind

Fore Print
(hind print is slightly smaller)
Length: 2.1–3 in (5.3–7.5 cm)
Width: 1.6–2.3 in (4–5.8 cm)

Straddle
2–3.5 in (5–9 cm)

Stride
Trotting: 12–18 in (30–45 cm)
Side-trotting: 14–21 in (36–53 cm)

Size (vixen is slightly smaller)
Height: 14 in (35 cm)
Length: 22–25 in (55–65 cm)

Weight
7–15 lb (3.2–7 kg)

trotting   *side-trotting*

RED FOX
Vulpes vulpes

This beautiful and notoriously cunning fox, found throughout the region, prefers mountainous forests and open areas. It is a very adaptable and intelligent animal.

The finer details of a Red Fox's tracks are blurred by its foot hairs, so only parts of the toes and heel pads show. The horizontal or slightly curved bar across the fore heel pad is diagnostic. A trotting Red Fox leaves a distinctive, straight trail of alternating prints—the hind print direct registers on the wider fore print. When a fox side-trots, it leaves print pairs in which the hind print falls to one side of the fore print in typical canid fashion. Foxes gallop like Coyotes (p. 30). The faster the gallop, the straighter the gallop group.

Similar Species: Other canid prints lack the bar across the fore heel pad. Gray Fox (p. 34) prints are smaller. Domestic Dog (*Canis familiaris*) prints can be of similar size. Small Coyote prints are similar, but they have a wider straddle, and the toe marks are more bulbous.

Gray Fox

fore

hind

Fore Print
(hind print is slightly smaller)
Length: 1.3–2.1 in (3.3–5.3 cm)
Width: 1.1–1.5 in (2.8–3.8 cm)
Straddle
2–4 in (5–10 cm)
Stride
Walking/Trotting: 10–12 in (25–30 cm)
Size
Height: 14 in (35 cm)
Length: 21–30 in (53–75 cm)
Weight
7–15 lb (3.2–7 kg)

walking

GRAY FOX
Urocyon cinereoargenteus

   This small, shy fox is widespread, but it especially prefers woodlands and chaparral country. The Gray Fox is the only fox that climbs trees, which it does either for safety or to forage.

   The forefoot registers better than the smaller hind foot, and the hind foot's long, semi-retractable claws do not always register. The heel pads are often unclear—they sometimes show up just as small, round dots. When it walks, this fox leaves a neat alternating track pattern; when it trots, its prints fall in pairs, with the fore print set diagonally behind the hind print. The Gray Fox's gallop group is like the Coyote's (p. 30).

Similar Species: The Red Fox (p. 32) has heel pads with a bar across them; in general, its prints are larger and less clear (because of thick fur), its stride is longer, and its straddle is narrower. Coyote tracks are much larger. Domestic Cat (p. 38) and Bobcat (p. 36) prints lack claw marks and have larger, less symmetrical heel pads.

Bobcat

fore

hind

Fore Print
(hind print is slightly smaller)
Length: 1.8–2.5 in (4.5–6.5 cm)
Width: 1.8–2.5 in (4.5–6.5 cm)

Straddle
4–7 in (10–18 cm)

Stride
Walking: 8–16 in (20–40 cm)
Running: 4–8 ft (1.2–2.4 m)

Size (female is slightly smaller)
Height: 20–22 in (50–55 cm)
Length: 25–30 in (65–75 cm)

Weight
15–35 lb (7–16 kg)

walking

ambling to loping

BOBCAT (Wildcat)

Lynx rufus

Widely distributed in the Appalachians, the Bobcat is a stealthy and usually nocturnal hunter, so it is seldom seen. Very adaptable, it can leave tracks anywhere from wild mountainsides to chaparral and even into residential areas.

When a Bobcat walks, its hind feet usually register directly on the larger fore prints. As a Bobcat picks up speed, its trail becomes an ambling pattern of paired prints, the hind leading the fore. At even greater speeds, it leaves four-print groups in a lope pattern. Especially the fore prints show asymmetry. The front part of the heel pad has two lobes and the rear part has three. In deep snow the Bobcat's feet leave draglines. The Bobcat marks its territory with half-buried scat along its meandering trail.

Similar Species: Large Domestic Cats (p. 38) have similar prints with a shorter stride and a narrower straddle. Coyote (p. 30), Fox (pp. 32–35) and Domestic Dog (*Canis familiaris*) tracks are narrower than they are long and show claw marks, and the fronts of their footpads are once-lobed; wild canid trails do not meander. Some mustelids' (pp. 46–55) four-toed hind prints may also seem similar.

Domestic Cat

fore

hind

Fore Print
(hind print is slightly smaller)
Length: 1–1.6 in (2.5–4 cm)
Width: 1–1.8 in (2.5–4.5 cm)

Straddle
2.4–4.5 in (6–11 cm)

Stride
Walking: 5–8 in (13–20 cm)
Loping/Galloping:
    14–32 in (35–80 cm)

Size (male>female)
Height: 20–22 in (50–55 cm)
Length with tail: 30 in (75 cm)

Weight
6.5–13 lb (3–6 kg)

walking

loping to
galloping

DOMESTIC CAT
(House Cat)
Felis catus

The tracks of the familiar and abundant Domestic Cat can show up almost any place where there are people. Abandoned cats may roam farther afield, and these 'feral cats' lead a pretty wild and independent existence. Domestic Cats can come in many shapes sizes and colors.

As with all felines, the Domestic Cat's fore print and slightly smaller hind print both show four toe pads. Its retractable claws, kept clean and sharp for catching prey, do not register. Cat prints usually show a slight asymmetry, with one toe leading the others. A Domestic Cat makes a neat alternating walking track pattern, usually in direct register, as one would expect from this animal's fastidious nature. When a cat picks up speed it leaves clusters of four prints, the hind feet registering in front of the forefeet.

Similar Species: A small Bobcat (p. 36) may leave tracks similar to a very large Domestic Cat's. Fox (pp. 32–35) and Domestic Dog (*Canis familiaris*) prints show claw marks.

Raccoon

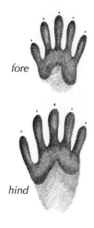

fore

hind

Fore Print
Length: 2–3 in (5–7.5 cm)
Width: 1.8–2.5 in (4.5–6.5 cm)

Hind Print
Length: 2.4–3.8 in (6–9.5 cm)
Width: 2–2.5 in (5–6.5 cm)

Straddle
3.3–6 in (8.5–15 cm)

Stride
Walking: 8–18 in (20–45 cm)
Bounding: 15–25 in (38–65 cm)

Size (female is slightly smaller)
Length: 24–37 in (60–95 cm)

Weight
11–35 lb (5–16 kg)

walking

bounding group

RACCOON
Procyon lotor

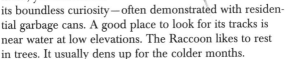

The inquisitive Raccoon, common in the Appalachians, is adored by some people for its distinctive face mask, yet disliked for its boundless curiosity—often demonstrated with residential garbage cans. A good place to look for its tracks is near water at low elevations. The Raccoon likes to rest in trees. It usually dens up for the colder months.

The Raccoon's unusual print, which looks like a human handprint, shows five well-formed toes; the small claws make dots. Its highly dexterous forefeet rarely leave heel prints, but its hind prints, which are generally much clearer, do show heels. The Raccoon's peculiar walking track pattern shows the left fore print next to the right hind print (or just in front) and vice versa. When a Raccoon is out in deep snow (which is rare), it may use a direct-registering walk. Raccoons occasionally bound, leaving clusters with hind prints in front of fore prints.

Similar Species: Unclear Opossum (p. 42) prints may look similar, but the Opossum drags its tail. In deep snow, Fisher (p. 46) or River Otter (p. 44) tracks may look similar.

Opossum

fore

hind

Fore Print
Length: 2–2.3 in (5–5.8 cm)
Width: 2–2.3 in (5–5.8 cm)

Hind Print
Length: 2.5–3 in (6.5–7.5 cm)
Width: 2–3 in (5–7.5 cm)

Straddle
4–5 in (10–13 cm)

Stride
Walking: 5–11 in (13–28 cm)

Size
Length: 2–2.5 ft (60–75 cm)

Weight
9–13 lb (4–6 kg)

walking

fast walking

OPOSSUM
Didelphis virginiana

This slow-moving, nocturnal marsupial is found throughout the Appalachians. Though it occupies many habitats, it prefers open woodland or brushland around waterbodies. It is quite tolerant of residential areas. The Opossum's tracks can often be seen in mud near the water and in snow during the warmer months of winter, but it tends to den up during severe weather.

The Opossum is an excellent climber, so its trail may lead to a tree. It has two walking habits: the common alternating pattern, with the hind prints registering on the fore prints, and a Raccoon-like (p. 40) paired-print pattern, with each hind print next to the opposing fore print. The very distinctive, long, inward-pointing thumb of the hind foot does not make a claw mark. In snow the dragline of the long, naked tail may be stained with blood; this thinly haired animal is not well adapted to cold and frequently suffers frostbite.

Similar Species: Prints in which the distinctive thumbs do not show, as in sand or fine snow, may be mistaken for a Raccoon's.

River Otter

fore

hind

Fore Print
Length: 2.5–3.5 in (6.5–9 cm)
Width: 2–3 in (5–7.5 cm)

Hind Print
Length: 3–4 in (7.5–10 cm)
Width: 2.3–3.3 in (5.8–8.5 cm)

Straddle
4–9 in (10–23 cm)

Stride
Loping: 12–27 in (30–70 cm)

Size
(female is two-thirds the size of male)
Length with tail: 3–4.3 ft (90–130 cm)

Weight
10–25 lb (4.5–11 kg)

loping (fast)

RIVER OTTER
Lontra canadensis

No animal knows how to have more fun than a River Otter. If you are lucky enough to watch one at play, you will not soon forget the experience. Widespread and well-adapted for the aquatic environment, this otter lives along waterbodies; an otter in the forest is usually on its way to another waterbody. Expect to see a wealth of evidence of an otter's presence along the riverbanks in its home territory.

In soft mud, the River Otter's five-toed feet, especially the hind ones, register evidence of webbing. The inner toes are set slightly apart. If the forefoot's metacarpal pad registers, it lengthens the print. Very variable, otter trails usually show a typical mustelid 2×2 loping. However, with faster gaits they show groups of four and three prints. The thick, heavy tail often leaves a dragline. This otter loves to slide in snow, often down riverbanks, leaving troughs nearly 1 foot (30 cm) wide. In summer it rolls and slides on the grass and in the mud.

Similar Species: Other mustelid trails lack conspicuous tail drag. The Marten (p. 48) has similar-sized prints. The Fisher (p. 46) usually has hairy feet, with the forefeet larger than the hind feet. Mink (p. 50) prints are about half the size.

Fisher

Fore Print
Length: 2.1–4 in (5.3–10 cm)
Width: 2.1–3.3 in (5.3–8.5 cm)

Hind Print
Length: 2.1–3 in (5.3–7.5 cm)
Width: 2–3 in (5–7.5 cm)

Straddle
3–7 in (7.5–18 cm)

Stride
Walking: 7–14 in (18–35 cm)
2x2 loping: 1–4.3 ft (30–130 cm)
Loping: 1–3 ft (30–90 cm)

Size (male>female)
Length with tail:
   34–40 in (85–100 cm)

Weight
3–12 lb (1.4–5.5 kg)

walking   *2x2 loping*

FISHER (Black Cat)
Martes pennanti

Once common in the Appalachians, the Fisher has declined in numbers and is only found in isolated populations in the northern ranges. This agile hunter is comfortable both on the ground and in the trees of mixed hardwood forests. It is one of the few predators to kill and eat Porcupines (p. 64)—if there are none, rabbits and hares (pp. 60–63) are its primary prey.

Though all five toes may register, the small inner toe frequently does not. Only the forefoot has a small heel pad that can show up in the print. The Fisher occasionally walks, making a direct-registering alternating track pattern, but it more often 2×2 lopes in typical mustelid fashion, leaving angled print pairs with the hind print direct registered on the fore print. Loping, its most common gait, produces three- and four-print groups (see the River Otter, p. 44). The patterns often vary within a short distance. The Fisher is not associated with water—it has been misnamed!

Similar Species: Male Marten (p. 48) tracks may resemble a small female Fisher's, but Martens weigh less and leave shallower prints. Otters have larger hind feet than forefeet. Other mustelids (pp. 50–55) have smaller prints. Four-toed Fisher prints may look like a Bobcat's (p. 36).

Marten

Fore and Hind Prints
Length: 1.8–2.5 in (4.5–6.5 cm)
Width: 1.5–2.8 in (3.8–7 cm)

Straddle
2.5–4 in (6.5–10 cm)

Stride
Walking: 4–9 in (10–23 cm)
Loping: 9–46 in (23–120 cm)

Size (male>female)
Length with tail:
    21–28 in (53–70 cm)

Weight
1.5–2.8 lb (0.7–1.3 kg)

walking | *2×2 loping*

MARTEN
(American Sable)
Martes americana

    This aggressive predator, once found in much of the northern Appalachians, has declined in numbers and is now only found in small, isolated populations.

    The Marten seldom leaves a clear print: often just four toes register, and the heel pad is undeveloped. In winter the hairiness of the feet often blurs all pad detail, especially the poorly developed palm pads. In the Marten's alternating walk, the hind feet register on the fore prints. In 2×2 loping, the hind prints fall on the fore prints to form slightly angled print pairs in a typical mustelid pattern. Its loping track patterns may appear as three- or four-print clusters (see the River Otter, p. 44). Follow the criss-crossing trails—if a Marten has scrambled up a tree, look for a sitzmark where it has jumped down.

Similar Species: Size and habitat are often key to distinguishing Marten, Fisher (p. 46) and Mink (p. 50) tracks. Female Fisher prints may resemble a large male Marten's but will be clearer. Male Mink prints overlap in size with small female Marten prints, but Mink rarely climb trees and (unlike Martens) are usually found near water.

Mink

fore

hind

Fore and Hind Prints
Length: 1.3–2 in (3.3–5 cm)
Width: 1.3–1.8 in (3.3–4.5 cm)

Straddle
2.1–3.5 in (5.3–9 cm)

Stride
Walking/Loping: 8–36 in (20–90 cm)

Size (male>female)
Length with tail: 19–28 in (48–70 cm)

Weight
1.5–3.5 lb (0.7–1.6 kg)

2×2 loping

MINK
Mustela vison

The lustrous Mink, widespread throughout the Appalachians, prefers watery habitats surrounded by brush or forest. At home as much on land as in water, this nocturnal hunter can be exciting to track. The Mink sometimes slides in snow like the River Otter (p. 44), carving out a trough up to 6 inches (15 cm) wide for an observant tracker to spot.

The Mink's fore print shows five (perhaps four) toes and five loosely connected palm pads in an arc, but the hind print shows only four palm pads. The metacarpal pad of the forefoot rarely registers, but the furred heel of the hind foot may register, lengthening the hind print. The Mink prefers the typical mustelid 2×2 loping, making consistently spaced, slightly angled double prints. Its diverse track patterns also include alternating walking, loping with three- and four-print groups (like the River Otter) and bounding (like rabbits and hares, pp. 60–63).

Similar Species: Small Martens (p. 48) may have similar prints, but without a consistent 2×2 loping gait, and they do not live near water. Weasels (pp. 52–55) make similar but smaller tracks. Small River Otter tracks (may overlap in size) often show webbing and lack a consistent double-print pattern.

Long-tailed Weasel

Fore and Hind Prints
Length: 1.1–1.8 in (2.8–4.5 cm)
Width: 0.8–1 in (2–2.5 cm)
Straddle
1.8–2.8 in (4.5–7 cm)
Stride
Loping: 9.5–43 in (24–110 cm)
Size (male>female)
Length with tail: 12–22 in (30–55 cm)
Weight
3–12 oz (85–340 g)

2x2 loping

LONG-TAILED WEASEL
Mustela frenata

Weasels are active year-round hunters with an avid appetite for rodents. The Long-tailed Weasel is the largest and most widely distributed of the two weasels in the Appalachians.

Following a weasel's tracks can reveal much about the nimble creature's activities. Tracks are most evident in winter, when weasels frequently burrow into the snow or pursue rodents into their holes. Some weasel trails may lead you up a tree. Weasels sometimes take to water. To identify the weasel species, pay close attention to the straddle, stride and loping patterns, and note the distribution and habitat. The usual weasel gait is a 2×2 lope, leaving a trail of paired prints. A weasel's light weight and small, hairy feet result in pad detail that is often unclear, especially in snow. Even with clear tracks, the inner (fifth) toe rarely registers.

The Long-tailed Weasel's typical 2×2 lope shows an irregular stride—sometimes short and sometimes long—with no consistent behavior. Like the Mink (p. 50), this weasel may bound like a rabbit or hare (pp. 60–63).

Similar Species: The Least Weasel (p. 54) shares some habitats with the Long-tailed Weasel, but the latter (except juveniles) has much larger tracks.

Least Weasel

Fore and Hind Prints
Length: 0.5–0.8 in (1.3–2 cm)
Width: 0.4–0.5 in (1–1.3 cm)
Straddle
0.8–1.5 in (2–3.8 cm)
Stride
Loping: 5–20 in (13–50 cm)
Size (male>female)
Length with tail: 6.5–9 in (17–23 cm)
Weight
1.3–2.3 oz (37–65 g)

2×2 loping

LEAST WEASEL
Mustela nivalis

The Least Weasel, the smallest of the two weasels in the Appalachians, is quite rare over most of its range.

The usual gait for any weasel is a 2×2 lope that results in a trail of paired prints. A weasel's light weight and small, hairy feet mean that the pad detail is often unclear, especially in snow. Even with clear tracks, the inner (fifth) toe rarely registers.

The Least Weasel makes the least-clear tracks. Look for its tracks around wetlands and in open woodlands and fields.

Similar Species: The Long-tailed Weasel (p. 52), except for juveniles, makes much larger tracks. Also, it does not frequent wet areas, preferring upland areas and woodlands.

Striped Skunk

fore

hind

Fore Print
Length: 1.5–2.2 in (3.8–5.6 cm)
Width: 1–1.5 in (2.5–3.8 cm)

Hind Print
Length: 1.5–2.5 in (3.8–6.5 cm)
Width: 1–1.5 in (2.5–3.8 cm)

Straddle
2.8–4.5 in (7–11 cm)

Stride
Walking/Bounding:
    2.5–8 in (6.5–20 cm)

Size
Length with tail:
    20–32 in (50–80 cm)

Weight
6–14 lb (2.7–6.5 kg)

walking (fast)   *bounding*

STRIPED SKUNK
Mephitis mephitis

This striking skunk is notorious for its vile smell—the lingering odor is often the best sign of its presence. Widespread throughout the Appalachians in a diversity of habitats, it prefers lower elevations. The Striped Skunk dens up in winter, coming out on warmer days and in spring.

Each foot has five toes. The smooth palm pads and small heel pads leave surprisingly small prints. The long claws on the forefeet often register. A skunk mostly walks—with such a potent smell for its defense, and those memorable black and white stripes, it rarely needs to run. This skunk's trail rarely shows any consistent pattern, but an alternating walking pattern may be evident. The greater a skunk's speed, the more the hind foot oversteps the fore. If it runs, its trail consists of clumsy, closely set four-print groups. In snow it drags its feet.

Similar Species: The Eastern Spotted Skunk (p. 58) makes smaller prints in a very random pattern. Mustelid (pp. 44–55) tracks are farther apart than those resulting from a skunk's shuffling gait, and this skunk's prints do not overlap.

Eastern Spotted Skunk

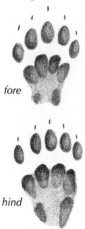

fore

hind

Fore Print
Length: 1–1.3 in (2.5–3.3 cm)
Width: 0.9–1.1 in (2.3–2.8 cm)

Hind Print
Length: 1.2–1.5 in (3–3.8 cm)
Width: 0.9–1.1 in (2.3–2.8 cm)

Straddle
2–3 in (5–7.5 cm)

Stride
Walking: 1.5–3 in (3.8–7.5 cm)
Bounding: 6–12 in (15–30 cm)

Size
Length: 13–25 in (33–65 cm)

Weight
0.6–2.2 lb (0.3–1 kg)

walking   bounding

EASTERN SPOTTED SKUNK
Spilogale putorius

This beautifully marked skunk, smaller than its striped cousin (the Striped Skunk, p. 56), is found throughout most of the Appalachians. It enjoys diverse habitats—such as scrubland, forests and farmland—but it is a rare sight because of its nocturnal habits, and because it dens up in winter, coming out only on warmer nights.

This skunk leaves a very haphazard trail as it forages for food on the ground. Occasionally it climbs trees, which it does with ease. The long claws on the forefeet often register, and the palm and heel may leave defined pad marks. Although this skunk rarely runs, when it does so it may bound along, leaving groups of four prints, hind in front of fore. It sprays only when truly provoked, so its powerful odor is less frequently detected than that of the Striped Skunk.

Similar Species: The Striped Skunk has larger prints and less scattered tracks with a shorter running stride (or it jumps), and it does not climb trees.

Snowshoe Hare

hind

fore

Fore Print
Length: 2–3 in (5–7.5 cm)
Width: 1.5–2 in (3.8–5 cm)
Hind Print
Length: 4–6 in (10–15 cm)
Width: 2–3.5 in (5–9 cm)
Straddle
6–8 in (15–20 cm)
Stride
Hopping: 0.8–4.3 ft (25–130 cm)
Size
Length: 12–21 in (30–53 cm)
Weight
2–4 lb (0.9–1.8 kg)

hopping

SNOWSHOE HARE
(Varying Hare)
Lepus americanus

This hare is well known for its color change from summer brown to winter white and for its huge hind feet, which enable it to 'float' on top of snow. Found only in the northern parts of the Appalachians, it frequents brushy areas in forests, which provide good cover from the Bobcat (p. 36) and the Coyote (p. 30), its most likely predators. Hares are most active at night.

As with other hares and rabbits, the Snowshoe Hare's most common track pattern is a hopping one, with triangular four-print groups; they can be quite long if the hare moves quickly. In winter, heavy fur on the hind feet (which are much larger than the forefeet) thickens the toes, which can splay out to further distribute the hare's weight on snow. Hares make well-worn runways that are often used as escape runs. You may encounter a resting hare, because hares do not live in burrows. Twigs and stems neatly cut at a 45° angle also indicate this hare's presence.

Similar Species: Eastern Cottontail (p. 62) prints are similar, but smaller.

61

Eastern Cottontail

fore

hind

Fore Print
Length: 1–1.5 in (2.5–3.8 cm)
Width: 0.8–1.3 in (2–3.3 cm)
Hind Print
Length: 3–3.5 in (7.5–9 cm)
Width: 1–1.5 in (2.5–3.8 cm)
Straddle
4–5 in (10–13 cm)
Stride
Hopping: 0.6–3 ft (18–90 cm)
Size
Length: 12–17 in (30–43 cm)
Weight
1.3–3 lb (0.6–1.4 kg)

hopping

EASTERN COTTONTAIL
Sylvilagus floridanus

Widespread throughout the Appalachians, this abundant rabbit prefers brushy areas in grasslands and cultivated areas. It might be found in dense vegetation, where it hides from predators such as the Bobcat (p. 36) and the Coyote (p. 30). Largely nocturnal, the Eastern Cottontail can be seen at dawn or dusk and on darker days.

Like other rabbits and hares, this cottontail most commonly leaves a triangular grouping of four prints, with the larger hind prints (which can appear pointed) falling in front of the fore prints (which may overlap). The hairiness of the toes blurs the pad detail in the prints. If you follow the trail of this rabbit, it might startle you if it flies out from its 'form,' a depression in the ground in which it rests.

Similar Species: The Allegheny Cottontail (*S. obscurus*) and the New England Cottontail (*S. transitionalis*) are both similar but less common. The Snowshoe Hare (p. 60) has larger prints, especially the hind ones. Tree squirrel (pp. 74–83) tracks show a similar pattern, but their fore prints are more consistently side by side.

Porcupine

fore

hind

Fore Print
Length: 2.3–3.3 in (5.8–8.5 cm)
Width: 1.3–1.9 in (3.3–4.8 cm)

Hind Print
Length: 2.8–4 in (7–10 cm)
Width: 1.5–2 in (3.8–5 cm)

Straddle
5.5–9 in (14–23 cm)

Stride
Walking: 5–10 in (13–25 cm)

Size
Length with tail: 25–40 in (65–100 cm)

Weight
10–28 lb (4.5–13 kg)

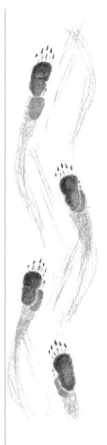

walking

PORCUPINE
Erethizon dorsatum

This notorious rodent rarely runs—its many long quills are a formidable defense. Found in the northern Appalachians, the Porcupine shows a preference for forests, but can also be seen in more open areas.

The Porcupine's preferred pigeon-toed, waddling gait leaves an alternating track pattern, with the hind print registered on or slightly in front of the shorter fore print. Look for long claw marks on all prints. The fore print shows four toes, and the hind print shows five. Clear prints may show the unusual pebbly surface of the solid heel pads, but a Porcupine's tracks are often scratch-marked by its heavy, spiny tail. In deeper snow this squat animal drags its feet, and it may leave a trough with its body. A Porcupine's trail might lead you to a tree, where this animal spends much of its time feeding; if so, look for chewed bark or nipped twigs on the ground.

Similar Species: The Porcupine's distinctive prints differ from those of any other animal in this region.

Beaver

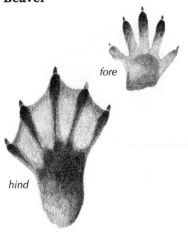

fore

hind

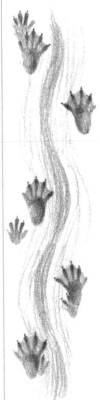

Fore Print
Length: 2.5–4 in (6.5–10 cm)
Width: 2–3.5 in (5–9 cm)
Hind Print
Length: 5–7 in (13–18 cm)
Width: 3.3–5.3 in (8.5–13 cm)
Straddle
6–11 in (15–28 cm)
Stride
Walking: 3–6.5 in (7.5–17 cm)
Size
Length with tail: 3–4 ft (90–120 cm)
Weight
28–75 lb (13–34 kg)

walking

BEAVER
Castor canadensis

Few animals leave as many signs of their presence as the Beaver, the largest North American rodent and a common sight around water. Look for the conspicuous dams and lodges—capable of changing the local landscape—and the stumps of felled trees. Check trunks gnawed clean of bark for marks of the Beaver's huge incisors. Scent mounds marked with castoreum, a strong-smelling yellowish fluid that Beavers produce, also indicate recent activity.

Rarely do all five toes on each foot register. Check the large hind prints for signs of webbing and broad toenails, though the nail of the second inner toe usually does not show. Irregular foot placement in the alternating walking gait may produce a direct register or a double register. The Beaver's thick, scaly tail may mar its tracks, as can the branches that it drags about for construction and food. Repeated path use results in well-worn trails.

Similar Species: The Beaver's many signs, including its large hind prints, minimize confusion. Muskrat (p. 68) prints are smaller.

Muskrat

fore

hind

Fore Print
Length: 1.1–1.5 in (2.8–3.8 cm)
Width: 1.1–1.5 in (2.8–3.8 cm)

Hind Print
Length: 1.6–3.2 in (4–8 cm)
Width: 1.5–2.1 in (3.8–5.3 cm)

Straddle
3–5 in (7.5–13 cm)

Stride
Walking: 3–5 in (7.5–13 cm)
Running: to 1 ft (30 cm)

Size
Length with tail: 16–25 in (40–65 cm)

Weight
2–4 lb (0.9–1.8 kg)

walking

MUSKRAT
Ondatra zibethicus

Like the Beaver (p. 66), this rodent is found throughout the Appalachians, wherever there is water. Beavers are very tolerant of Muskrats and even allow them to live in parts of their lodges. Active year-round, the Muskrat leaves plenty of signs. It digs extensive networks of burrows, often undermining riverbanks, so do not be surprised if you suddenly fall into a hidden hole! Also look for small lodges in the water and beds of vegetation on which the Muskrat rests, suns and feeds in summer.

The small inner toe of the five on each forefoot rarely registers. The hind print shows five well-formed toes that may have a 'shelf' around them, created by stiff hairs that aid in swimming. The common alternating walking pattern shows print pairs that alternate from side to side, with the hind print just behind or slightly overlapping the fore print. In snow, a Muskrat's feet drag, and its tail leaves a sweeping dragline.

Similar Species: Few animals share this water-loving rodent's habits. Beavers make larger tracks and leave many other signs.

Woodchuck

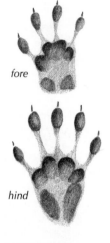

fore

hind

Fore and Hind Prints
Length: 1.8–2.8 in (4.5–7 cm)
Width: 1–2 in (2.5–5 cm)

Straddle
3.3–6 in (8.5–15 cm)

Stride
Walking: 2–6 in (5–15 cm)
Bounding: 6–14 in (15–35 cm)

Size (male>female)
Length with tail:
    20–25 in (50–65 cm)

Weight
5.5–12 lb (2.5–5.5 kg)

walking

bounding

WOODCHUCK
(Whistle Pig, Groundhog, Marmot)

Marmota monax

This robust member of the squirrel family is a common sight in open woodlands and adjacent open areas throughout this region. Always on the watch for predators, but not too troubled by humans, the Woodchuck never wanders far from its burrow. This marmot hibernates during winter and emerges in early spring; look for tracks in late spring snowfalls and in the mud around the burrow entrances.

A Woodchuck's fore print shows four toes, three palm pads and two heel pads (not always evident). The hind print shows five toes, four palm pads and two poorly registering heel pads. The Woodchuck usually leaves an alternating walking pattern, with the hind print registered on the fore print. When a Woodchuck runs from danger, it makes groups of four prints, hind ahead of fore.

Similar Species: A small Raccoon's (p. 40) bounding track pattern will be similar, but it will show five-toed fore prints.

Eastern Chipmunk

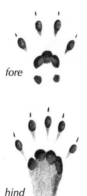

fore

hind

Fore Print
Length: 0.8–1 in (2–2.5 cm)
Width: 0.4–0.8 in (1–2 cm)

Hind Print
Length: 0.7–1.3 in (1.8–3.3 cm)
Width: 0.5–0.9 in (1.3–2.3 cm)

Straddle
2–3.2 in (5–8 cm)

Stride
Running: 7–15 in (18–38 cm)

Size
Length with tail: 7–10 in (18–25 cm)

Weight
2.5–5 oz (70–140 g)

bounding

EASTERN CHIPMUNK
Tamias striatus

Look for this delightful character throughout the Appalachians. The Eastern Chipmunk is found in a variety of habitats, from the dense forest floor to open areas near buildings. You are more likely to see or hear this rodent, which is highly active during summer, than to notice its tracks. This large chipmunk is happiest on the ground, but it will gladly climb sturdy oak trees to harvest juicy, ripe acorns. It enters a deep sleep in winter, waking up from time to time to have a meal.

Chipmunks are so light that their tracks rarely show fine details. The forefeet each have four toes and the hind feet have five. Chipmunks run on their toes, so the two heel pads of the forefeet seldom register; the hind feet have no heel pads. Their erratic track patterns, like those of many of their cousins, show the hind feet registered in front of the forefeet. A chipmunk trail often leads to extensive burrows.

Similar Species: No other chipmunks inhabit this region. Tree squirrels (pp. 74–83) usually have larger prints and a wider straddle, and they are more likely to make midwinter tracks. Mouse (pp. 94–97) tracks are smaller.

Eastern Gray Squirrel

fore

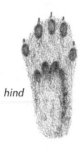

hind

Fore Print
Length: 1–1.8 in (2.5–4.5 cm)
Width: 1 in (2.5 cm)

Hind Print
Length: 2.3–3 in (5.8–7.5 cm)
Width: 1.1–1.5 in (2.8–3.8 cm)

Straddle
3.8–6 in (9.5–15 cm)

Stride
Bounding: 0.7–3 ft (22–90 cm)

Size
Length with tail: 17–20 in (43–50 cm)

Weight
14–25 oz (400–710 g)

bounding

EASTERN GRAY SQUIRREL

Sciurus carolinensis

This large, familiar squirrel can be a common sight in deciduous and mixed forests through-out this region, even in urban areas. Active all year, the Eastern Gray Squirrel can leave a wealth of evidence, especially in winter as it scurries about digging up nuts that it buried during the previous fall.

The Eastern Gray Squirrel leaves a typical squirrel track when it runs or bounds. The hind prints fall slightly in front of the fore prints. A clear fore print shows four toes with sharp claws, four fused palm pads and two heel pads. The hind print shows five toes and four palm pads; if the full heel-length registers, it also shows two small heel pads.

Similar Species: Fox Squirrel (p. 76) prints are as big or larger. Red Squirrel (p. 78) prints are smaller. Chipmunks (p. 72) and flying squirrels (pp. 80–83) make smaller tracks in a similar pattern but with narrower straddles. Rabbits and hares (pp. 60–63) make longer track patterns, and their forefeet rarely register side by side when they run.

Fox Squirrel

fore

hind

Fore Print
Length: 1–1.9 in (2.5–4.5 cm)
Width: 1–1.7 in (2.5–4.3 cm)
Hind Print
Length: 2–3.3 in (5–7.5 cm)
Width: 1.5–1.9 in (3.8–4.8 cm)
Straddle
4–6 in (10–15 cm)
Stride
Bounding: 0.7–3 ft (22–90 cm)
Size
Length with tail: 18–28 in (45–70 cm)
Weight
1–2.4 lb (0.5–1.1 kg)

bounding

FOX SQUIRREL
Sciurus niger

This squirrel is much like the Eastern Gray Squirrel (p. 74), but larger and with a yellowish underside. It can be a common sight in deciduous forests with plenty of nut trees and in open areas or woodlands throughout the Appalachians. Piles of nutshells at tree bases indicate its favorite feeding sites. Active all year, the Fox Squirrel spends a lot of time foraging on the ground, often collecting nuts that it buried singly during the previous fall.

A clear fore print shows four toes with claws evident, four fused palm pads and two heel pads. The hind print shows five toes, four palm pads and sometimes a heel. When it runs or bounds, the Fox Squirrel makes a typical squirrel track, the hind prints slightly in front of the fore prints, with the prints in each pair roughly side by side.

Similar Species: The Eastern Gray Squirrel generally leaves smaller prints. Chipmunk (p. 72) and flying squirrel (pp. 80–83) tracks are smaller and have narrower straddles. Red Squirrel (p. 78) prints are much smaller. A rabbit or hare (pp. 60–63) makes a longer print pattern, and its fore prints rarely register side by side when it runs.

Red Squirrel

fore

hind

Fore Print
Length: 0.8–1.5 in (2–3.8 cm)
Width: 0.5–1 in (1.3–2.5 cm)

Hind Print
Length: 1.5–2.3 in (3.8–5.8 cm)
Width: 0.8–1.3 in (2–3.3 cm)

Straddle
3–4.5 in (7.5–11 cm)

Stride
Bounding: 8–30 in (20–75 cm)

Size
Length with tail:
9–15 in (23–38 cm)

Weight
2–9 oz (55–260 g)

bounding

*bounding
(deep snow)*

RED SQUIRREL
(Pine Squirrel, Chickaree)
Tamiasciurus hudsonicus

When you enter a Red Squirrel's territory, the inhabitant greets you with a loud, chattering call. Another obvious sign of this forest dweller, which is found throughout most of the Appalachians, is its large middens—piles of cone scales and cores left beneath trees—that indicate favorite feeding sites.

Active year-round in their small territories, Red Squirrels leave an abundance of trails that lead from tree to tree or down a burrow. These energetic animals mostly bound, leaving groups of four prints, the hind prints in front of the fore prints, which tend to be side by side (but not always). Four toes show on each fore print, and five show on each hind print. The heels often do not register when squirrels move quickly. In deeper snow the prints merge to form pairs of diamond-shaped tracks.

Similar Species: The larger Eastern Gray Squirrel (p. 74) is also found throughout the region. The Eastern Chipmunk (p. 72) and flying squirrels (pp. 80–83) make tracks in a similar pattern, but they are smaller and have narrower straddles.

Northern Flying Squirrel

fore

hind

Fore Print
Length: 0.5–0.8 in (1.3–2 cm)
Width: 0.8 in (1.3 cm)

Hind Print
Length: 1.5–1.8 in (3.8–4.5 cm)
Width: 0.8 in (2 cm)

Straddle
3–3.8 in (7.5–9.5 cm)

Stride
Bounding: 11–30 in (28–75 cm)

Size
Length with tail: 11–15 in (28–38 cm)

Weight
4–6.5 oz (110–180 g)

*sitzmark into
bounding*

NORTHERN FLYING SQUIRREL

Glaucomys sabrinus

This soft-furred brown acrobat can be found in coniferous forests of the northern Appalachians. It prefers widely spaced forests, where it can glide from tree to tree by night, using the membranous flaps of skin between its forelegs and hind legs. Up to 10 Northern Flying Squirrels will den up together in a tree cavity for warmth in winter.

Because of its gliding, this squirrel does not leave as many tracks as most other squirrels do. Evidence is scarce in summer, but in winter you may come across a sitzmark (the distinctive pattern that it made where it landed in the snow) and a short bounding trail, which it made as it rushed off to the nearest tree or to do some quick foraging. The bounding track pattern is typical of squirrels and other rodents, but usually the hind feet register only slightly in front of the forefeet, and often all four feet register in a row.

Similar Species: The Southern Flying Squirrel (p. 82) typically has a different bounding pattern, and its tracks are similar but smaller. The Red Squirrel (p. 78) usually makes larger prints and rarely leaves a sitzmark, but unclear tracks in deep snow can be indistinguishable. Chipmunk (p. 72) prints are smaller, with a narrower straddle.

Southern Flying Squirrel

fore

hind

Fore Print
Length: 0.3–0.5 in (0.8–1.3 cm)
Width: 0.4 in (1 cm)

Hind Print
Length: 0.9–1.3 in (2.3–3.3 cm)
Width: 0.5 in (1.4 cm)

Straddle
2–2.5 in (5–6.5 cm)

Stride
Bounding: 7–22 in (18–55 cm)

Size
Length with tail: 8–10 in (20–25 cm)

Weight
1.5–3.2 oz (43–90 g)

bounding (in snow)

SOUTHERN FLYING SQUIRREL
Glaucomys volans

Like the much larger Northern Flying Squirrel (p. 80), this aerialist is capable of long-distance gliding using the membranes that connect its forelegs and hind legs. The Southern Flying Squirrel lives in pine-dominated hardwood forests and is primarily nocturnal.

This squirrel sometimes bounds in a manner similar to the Northern Flying Squirrel, but with a narrower straddle that does not exceed 3 inches (7.6 cm). However, information from Mark Elbroch, a tracking expert from New England, shows that the more characteristic pattern for the Southern Flying Squirrel is a bound where the front tracks register in front of the rear tracks, and the straddle is even narrower. Flying squirrels often come down trunks of trees, rather than gliding to the earth.

Similar Species: The Northern Flying Squirrel makes similar but larger sitzmarks and tracks. The Red Squirrel (p. 78) makes much larger prints and rarely leaves a sitzmark. Chipmunk (p. 72) prints are smaller, with a narrower straddle.

Appalachian Woodrat

fore

hind

Fore Print
Length: 0.6–0.8 in (1.5–2 cm)
Width: 0.4–0.5 in (1–1.3 cm)

Hind Print
Length: 1–1.5 in (2.5–3.8 cm)
Width: 0.6–0.8 in (1.5–2 cm)

Straddle
2.3–2.8 in (5.8–7 cm)

Stride
Walking: 1.8–3 in (4.5–7.5 cm)
Bounding: 5–8 in (13–20 cm)

Size
Length with tail: 14–17 in (35–43 cm)

Weight
13–16 oz (370–450 g)

walking   *bounding*

APPALACHIAN WOODRAT
(Allegheny Woodrat)
Neotoma magister

This nocturnal woodrat lives in the rocky areas in the central and eastern parts of the region, but it feeds on foliage, seeds, ferns and fungi and therefore requires nearby vegetation. The trail of the Appalachian Woodrat might lead you to a distinctive mass of a nest, most often in a crevice, shrub or burrow.

Four toes show on the fore print, and the hind shows five. The short claws rarely register. Woodrats often walk in an alternating fashion, with the hind print direct registering on the fore print. This woodrat frequently bounds, leaving a pattern of four prints, with the larger hind print in front of the diagonally placed fore prints. The stride tends to be short relative to the size of the prints.

Similar Species: The Eastern Woodrat (*N. floridana*) makes identical tracks. The Norway Rat (p. 86), with similar prints, is usually found close to human activity. Woodchuck (p. 70) prints are similar but much larger.

Norway Rat

fore

hind

Fore Print
Length: 0.7–0.8 in (1.8–2 cm)
Width: 0.5–0.7 in (1.3–1.8 cm)

Hind Print
Length: 1–1.3 in (2.5–3.3 cm)
Width: 0.8–1 in (2–2.5 cm)

Straddle
2–3 in (5–7.5 cm)

Stride
Walking: 1.5–3.5 in (3.8–9 cm)
Bounding: 9–20 in (23–50 cm)

Size
Length with tail: 13–19 in (33–48 cm)

Weight
7–18 oz (200–510 g)

walking

NORWAY RAT
(Brown Rat)
Rattus norvegicus

Active both day and night, this despised rat is widespread almost anywhere that humans have decided to build their homes. Not entirely dependent on people, it may live in the wild as well.

The fore print shows four toes, and the hind print shows five. When it bounds, this colonial rat leaves four-print groups, with the hind prints in front of the diagonally placed fore prints. Sometimes one of the hind feet direct registers on a fore print, creating a three-print group. This rat more commonly leaves an alternating walking pattern with the larger hind prints close to or overlapping the fore prints; the hind heel does not show. The tail often leaves a dragline in snow. Rats live in groups, so you may find many trails together, often leading to their 5-inch (2-cm) wide burrows.

Similar Species: Woodrat (p. 84) tracks may be similar, but woodrats rarely associate with human activity, except in abandoned buildings. The Black Rat (*R. rattus*), also widespread, leaves similar tracks. Mouse (pp. 94–97) prints are much smaller. Red Squirrel (p. 78) tracks show distinctive squirrel traits. Chipmunk (p. 72) tracks are smaller, and the gaits differ.

Hispid Cotton Rat

fore

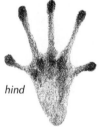

hind

Fore Print
Length: 0.5–0.7 in (1.3–1.8 cm)
Width: 0.5–0.7 in (1.3–1.8 cm)

Hind Print
Length: 0.6–1 in (1.5–2.5 cm)
Width: 0.6–0.8 in (1.5–2 cm)

Straddle
1.3–1.5 in (3.3–3.8 cm)

Stride
Walking; 1.3 in (3.3 cm)

Size
Length with tail: 8–14 in (20–35 cm)

Weight
2.8–7 oz (80–200 g)

walking

HISPID COTTON RAT
Sigmodon hispidus

The Hispid Cotton Rat, found only in the southern Appalachians, makes itself unpopular by eating valuable crops. Though keen on devouring almost anything green, it prefers grassy fields. It stays close to home, and consequently the rat's little runways clearly mark its routes to favored feeding sites.

The fore prints show four toes, but the larger hind prints usually show five. The heel of the hind foot will not always register, especially if the rat is moving fast. This medium-sized rodent leaves a typical walking track pattern in which the hind print registers on and slightly behind the fore print. Watch for its nests, which are balls of woven grass, and small piles of cut grass.

Similar Species: The widespread Marsh Rice Rat (p. 90) prefers marshes and leaves slightly smaller tracks. Woodrat (p. 84) tracks are similar but have a wider straddle. The Norway Rat (p. 86) also makes similar tracks, but it is usually found close to human activity.

Marsh Rice Rat

fore

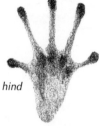

hind

Fore Print
Length: 0.5 in (1.3 cm)
Width: 0.5 in (1.3 cm)
Hind Print
Length: 0.6–0.9 in (1.5–2.3 cm)
Width: 0.4–0.8 in (1–2 cm)
Straddle
1.5 in (3.8 cm)
Stride
Walking: 1.2–2 in (3–5 cm)
Size
Length with tail: 7.5–12 in (19–30 cm)
Weight
1–2.8 oz (28–80 g)

walking

MARSH RICE RAT

Oryzomys palustris

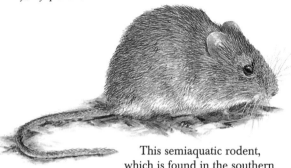

This semiaquatic rodent, which is found in the southern part of the Appalachian region, is known for its ability to dive and swim underwater, where it feeds on lush aquatic plants and invertebrates. Primarily nocturnal, it makes its home in the sedges or cattails of marshy areas. It sometimes damages nearby crops.

The considerable activity of this rice rat produces extensive runways amidst marsh vegetation. In dusty or muddy areas, clear hind prints will show five toes, and clear fore prints will show four. The rat's walking track pattern usually shows the hind print immediately behind the fore print, but the prints may direct register. If you follow the trail of a rice rat, it will likely lead to a softball-sized nest made of woven grass.

Similar Species: The Hispid Cotton Rat (p. 88) has very similar, slightly larger tracks, but it inhabits drier grassy areas. Woodrats (p. 84) and the Norway Rat (p. 86) are larger and usually do not live in such marshy areas.

Woodland Vole

fore

hind

Fore Print
Length: 0.5 in (1.3 cm)
Width: 0.5 in (1.3 cm)

Hind Print
Length: 0.6 in (1.5 cm)
Width: 0.5–0.8 in (1.3–2 cm)

Straddle
1.3–2 in (3.3–5 cm)

Stride
Walking/Trotting: 0.8 in (2 cm)
Bounding: 2–6 in (5–15 cm)

Size
Length with tail:
    4–5.5 in (10–14 cm)

Weight
0.8–1.3 oz (23–37 g)

walking

bounding (in snow)

WOODLAND VOLE
(Pine Vole)
Microtus pinetorum

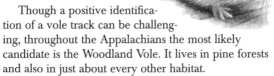

Though a positive identification of a vole track can be challenging, throughout the Appalachians the most likely candidate is the Woodland Vole. It lives in pine forests and also in just about every other habitat.

When clear (which is seldom), vole fore prints show four toes, and hind prints show five. A vole's walk and trot both leave a paired alternating track pattern with a hind print occasionally direct registered on a fore print. Voles usually opt for a faster bounding; the resulting print pairs show the hind prints registered on the fore prints. This vole lopes quickly across open areas, creating a three-print track pattern. Voles stay under the snow in winter; when it melts, look for distinctive piles of cut grass from their ground nests. The bark at the bases of shrubs may show tiny teeth marks left by gnawing. In summer, well-used vole paths appear as little runways in the grass.

Similar Species: The Southern Red-backed Vole (*Cletherionomys gapperi*) and the Rock Vole (*M. chrotorrhinus*) are also common here. Mouse (pp. 94–97) bounding tracks show four-print groups.

Cotton Mouse

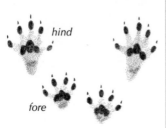

hind

fore

bounding group

Fore Print
Length: 0.3 in (0.8 cm)
Width: 0.3 in (0.8 cm)
Hind Print
Length: 0.6 in (1.5 cm)
Width: 0.4 in (1 cm)
Straddle
1.4–1.8 in (3.6–4.5 cm)
Stride
Bounding: 2–5 in (5–13 cm)
Size
Length with tail:
   6–12 in (15–30 cm)
Weight
0.7–1.6 oz (20–45 g)

bounding

*bounding
(in snow)*

COTTON MOUSE
Peromyscus gossypinus

   The nocturnal and seldom-seen Cotton Mouse
is widespread throughout the southern Appalachians.
It prefers swampland, but it also frequents forests, rocky
areas and even beaches. Its tracks may lead you up
a tree, down a burrow or into a river—this mouse is
a capable swimmer.

   A Cotton Mouse fore print shows four toes, three
palm pads and two heel pads. A hind print shows five
toes (the fifth is often unclear) and three palm pads; the
heel pads rarely register. Bounding tracks show hind
prints in front of close-set fore prints. In soft snow or
sand the prints may merge and appear as larger pairs,
and tail drag will be evident.

Similar Species: Many less common species of mice have
near-identical tracks. The House Mouse's (*Mus musculus*)
tracks are very similar, but the House Mouse associates
more with humans. Voles (p. 92) tend to trot and have a
longer stride. Jumping mouse (p. 96) prints may be similar
in size, but show long, thin toes. Chipmunks (p. 72) have a
wider straddle. Shrews (p. 98) have a narrower straddle.

Meadow Jumping Mouse

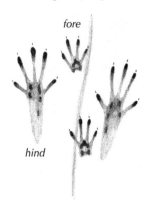

fore

hind

bounding

Fore Print
Length: 0.3–0.5 in (0.8–1.3 cm)
Width: 0.3–0.5 in (0.8–1.3 cm)

Hind Print
Length: 0.5–1.3 in (1.3–3.3 cm)
Width: 0.5–0.7 in (1.3–1.8 cm)

Straddle
1.8–1.9 in (4.5–4.8 cm)

Stride
Bounding: 7–18 in (18–45 cm)
In alarm: 3–6 ft (90–180 cm)

Size
Length with tail: 7–9 in (18–23 cm)

Weight
0.6–1.3 oz (17–35 g)

MEADOW JUMPING MOUSE
Zapus hudsonius

Congratulations if you find and successfully identify the tracks of the Meadow Jumping Mouse! Though it lives throughout the Appalachian region, its preference for grassy meadows and its long, deep winter hibernation (about six months!) make locating tracks very difficult.

Jumping mouse tracks are distinctive if you do find them. The two smaller forefeet register between the long hind feet; the long heels do not always register, and some prints show just the three long middle toes. The toes on the forefeet may splay so much that the side toes point backward. When they bound, jumping mice make short leaps. The tail may leave a dragline in soft mud or unseasonable snow. Clusters of cut grass stems, about 5 inches (13 cm) long, lying in meadows are a more abundant sign of this rodent.

Similar Species: The Woodland Jumping Mouse (*Napaeozapus insignis*), with similar tracks, can also be found in the mountains. Cotton Mouse (p. 94) tracks may have the same straddle. Heel-less hind prints may be mistaken for a vole's (p. 92) or a small bird's (p. 126) or an amphibian's (pp. 128–133).

Masked Shrew

hind

fore

bounding group

Fore Print
Length: 0.2 in (0.5 cm)
Width: 0.2 in (0.5 cm)
Hind Print
Length: 0.6 in (1.5 cm)
Width: 0.3 in (0.8 cm)
Straddle
0.8–1.3 in (2–3.3 cm)
Stride
Bounding: 1.2–3 in (3–7.5 cm)
Size
Length with tail: 2.3–4.5 in (7–11 cm)
Weight
0.1–0.3 oz (3–9 g)

bounding

MASKED SHREW

Sorex cinereus

Though there are several species of tiny, frenetic shrews in the region, the widespread and adaptable Masked Shrew is a likely candidate if you find tracks. This shrew is just as happy in woodlands as it is around open marshes. Its rapid activity makes it difficult to observe closely.

In its energetic and unending quest for food, a shrew usually leaves a four-print bounding pattern, but it may slow to an alternating walking gait. The individual prints in a group are often indistinct, but in mud or shallow, wet snow you can even count the five toes on each print. In deeper snow a shrew's tail often leaves a dragline. If a shrew tunnels under the snow, it may leave a snow ridge on the surface. You may find that a shrew's trail disappears down a burrow.

Similar Species: Other shrews in the area, all with similar tracks, include the Smoky Shrew (*S. fumeus*), the Southeastern Shrew (*S. longirostris*), the Short-tailed Shrews (*Blarina brevicauda* and *B. carolinensis*) and the Least Shrew (*Cryptotis parva*). Mouse (pp. 94–97) fore prints show just four toes.

Eastern Mole

a molehill of the Eastern Mole

some molehills and ridges of the Eastern Mole

Size
Length: 4.5–6.5 in (11–17 cm)
Tail length: 1–1.5 in (2.5–3.8 cm)
Weight
2.5–5 oz (70–140 g)

EASTERN MOLE
Scalopus aquaticus

This soft-furred resident of the underworld is the
mole that you will most likely encounter in this region.
It can certainly leave a wealth of evidence that indicates
its presence, usually in pastures or open woodlands,
especially where the soil is light and moist and easy
to burrow in.

Moles, which seldom emerge from their subterranean
environment, create an extensive network of burrows
through which they forage. These burrows are sometimes
marked by ridges on the surface, though most of us are
more familiar with the hills that form where the mole
gets rid of excess soil from its burrows and for which
moles are frequently considered to be pests when they
mess up fine lawns. When rains moisten the soil and
bring worms to the surface, it can be entertaining to
watch the earth twitch and rise up as the mole satisfies
its voracious appetite.

Similar Species: The Hairy-tailed Mole (*Parascalops
breweri*), also found in the mountains, leaves similar
signs.

BIRDS, AMPHIBIANS & REPTILES

A guide to the animal tracks of the Appalachians is not complete without some consideration of the birds and amphibians found in the region.

Several bird species have been chosen to represent the main types common to this area. Remember, however, that individual bird species are not easily identified by track alone. Bird tracks can often be found in abundance in snow and are clearest in shallow, wet snow. The shores of streams and lakes are very reliable locations in which to find bird tracks—the mud there can hold a clear print for a long time. The sheer number of tracks made by shorebirds and waterfowl can be astonishing. Though some bird species prefer to perch in trees or soar across the sky, it can be entertaining to follow the tracks of birds that spend a lot of time on the ground. They can spin around in circles and lead you in all directions. The trail may suddenly end as the bird takes flight, or it might terminate in a pile of feathers, the bird having fallen victim to a hungry predator.

Many amphibians and turtles depend on moist environments, so look in the soft mud along the shores of lakes and ponds for their distinctive tracks. Though you may be able to distinguish frog tracks from toad tracks, because these amphibians generally move differently, it can be very difficult to identify the species. In drier environments, reptiles, which thrive in dryness, outnumber the amphibians, but they seldom leave good tracks.

Canada Goose

Print
Length: 4–5 in (10–13 cm)
Straddle
5–7 in (13–18 cm)
Stride
Walking: 5–7 in (13–18 cm)
Size
2.7–4 ft (80–120 cm)

CANADA GOOSE
Branta canadensis

This handsome goose is most likely to be encountered in open areas by lakes and ponds throughout the Appalachians. Its huge, webbed feet leave prints that can often be seen in abundance along the muddy shores of just about any waterbody, including those in urban parks, where the Canada Goose's green-and-white droppings can accumulate in prolific amounts.

There are three long toes, all facing forward, on each webbed foot. These toes register well, but the webbing between them does not always show in the print. The feet point inward, giving the prints a pigeon-toed appearance and perhaps accounting for the bird's waddling gait.

Similar Species: No other geese are common in the Appalachians. Many other waterfowl, such as ducks (p. 106), as well as gulls (p. 108), leave similar but usually smaller prints. Exceptionally large prints are probably a swan's (*Cygnus* spp.).

Mallard

Print
Length: 2–2.5 in (5–6.5 cm)
Straddle
4 in (10 cm)
Stride
to 4 in (10 cm)
Size
23 in (58 cm)

MALLARD
Anas platyrhynchos

male

female

This dabbling duck can be found on or near almost any waterbody in the Appalachians. The male, with its striking green head, is a familiar sight. However, at the end of summer, when the male's brilliant plumage turns dull, it may be mistaken for a Mallard hen. The Mallard is quite tolerant of humans. Wherever you find tracks you have a good chance of seeing the duck as well.

The webbed foot of the Mallard has three long toes that all point forward. Though the toes register well, the webbing between the toes does not always show in the print. The Mallard's inward-pointing feet give it a pigeon-toed appearance and perhaps account for its waddling gait, a characteristic for which ducks are known.

Similar Species: Other dabbling ducks and many gulls (p. 108) have similar prints. Exceptionally large prints are from a Canada Goose (p. 104) or a swan (*Cygnus* spp.).

Herring Gull

Print
Length: 3.5 in (9 cm)
Straddle
4–6 in (10–15 cm)
Stride
4.5 in (11 cm)
Size
Length: 23–25 in (58–65 cm)

HERRING GULL
Larus argentatus

The Herring Gull, with its long wings and webbed toes, is a strong long-distance flier as well as an excellent swimmer. Common in a variety of habitats, it can be found concentrated in great numbers around food sources such as garbage dumps and campgrounds.

Gulls leave slightly asymmetrical tracks that show three toes. They have claws that register outside the webbing, and the claw marks are usually attached to the footprint. Most gulls have quite a swagger to their gait, and they leave a trail with the tracks turned strongly inward.

Similar Species: Gull and duck (p. 106) species cannot be reliably distinguished by track alone, but smaller species have conspicuously smaller tracks.

Great Blue Heron

Print
Length: to 6.5 in (17 cm)
Straddle
8 in (20 cm)
Stride
9 in (23 cm)
Size
4.2–4.5 ft (1.3–1.4 m)

GREAT BLUE HERON
Ardea herodias

The refined and graceful image of this large heron symbolizes the precious wetlands in which it patiently hunts for food. Usually still and statuesque as it waits for a meal to swim by, the Great Blue Heron will have cause to walk from time to time, perhaps to find a better hunting location. Look for its large, slender tracks along the banks or mudflats of waterbodies.

Not surprisingly, a bird that lives and hunts with such precision walks in a similar fashion, leaving straight tracks that fall in a nearly straight line. Look for the slender rear toe in the print.

Similar Species: Other herons have similar prints, varying in size relative to the size of the bird.

Common Snipe

Print
Length: 1.5 in (3.8 cm)
Straddle
to 1.8 in (4.5 cm)
Stride
to 1.3 in (3.3 cm)
Size
11–12 in (28–30 cm)

COMMON SNIPE
Gallinago gallinago

This short-legged character is a resident of marshes and bogs, where its neat prints can often be seen in mud. Snipes are quite secretive when on the ground, and so you may be surprised if one suddenly flushes out from beneath your feet. If there is a Common Snipe in the air, you may hear an eerie whistle if it dives from the sky.

The Common Snipe's neat prints show four toes, including a small rear toe that points inward. The bird's short legs and stocky body give it a very short stride.

Similar Species: Many shorebirds, including the Spotted Sandpiper (p. 114), leave similar tracks.

Spotted Sandpiper

Print
Length: 0.8–1.3 in (2–3.3 cm)

Straddle
to 1.5 in (3.8 cm)

Stride
Erratic

Size
7–8 in (18–20 cm)

SPOTTED SANDPIPER
Actitis macularia

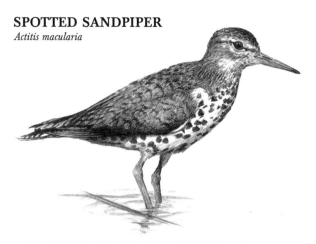

The bobbing tail of the Spotted Sandpiper is a common sight on the shores of lakes, rivers and streams, but you will usually find just one of these territorial birds in any given location. Because of its excellent camouflage, likely the first that you will see of this bird will be when it flies away, its fluttering wings close to the surface of the water.

As a sandpiper teeters up and down on the shore, it leaves trails of three-toed prints. Its fourth toe is very small and faces off to one side at an angle. Sandpiper tracks can have an erratic stride.

Similar Species: All sandpipers and plovers, including the Killdeer (*Charadrius vociferus*), leave similar tracks, although there is much diversity in size. The Common Snipe (p. 112) makes larger tracks.

Ruffed Grouse

Print
Length: 2–3 in (5–7.5 cm)
Straddle
0.5–2 in (1.3–5 cm)
Stride
Walking: 3–6 in (7.5–15 cm)
Size
15–19 in (38–48 cm)

RUFFED GROUSE
Bonasa umbellus

This ground-dweller prefers the quiet seclusion of coniferous forests, where its excellent camouflage usually affords it good protection. Although you might possibly find its tracks in mud, snow much improves your chances of finding them. If you follow a Ruffed Grouse trail quietly, you may be startled when the bird bursts from cover almost beneath your feet.

The three thick front toes leave very clear impressions, but the short rear toe, which is angled off to one side, does not always show up so well. This bird's neat, straight trail appears to reflect its cautious approach to life on the forest floor.

Similar Species: The Northern Bobwhite (*Colinus virginianus*) makes similar, but smaller prints.

Bald Eagle

Print
Length: 6 in (15 cm)
Straddle
to 10 in (25 cm)
Stride
Walking: to 9 in (23 cm)
Size
32–37 in (80–95 cm)

BALD EAGLE
Haliaeetus leucocephalus

The magnificent Bald Eagle, with its striking white head and brilliant yellow bill, is a familiar image to most of us. This predator is a scattered but year-round resident in much of this region. It prefers coasts, rivers and lakes, where it feeds largely on fish. Once persecuted by hunters, this icon of America is making a gradual comeback.

The track of a Bald Eagle is large and robust. Three toes extend forward, and one extends to the rear. The impressive, sharp talons register distinctly. If you find the remnants of fish or other carrion, look closely for Bald Eagle tracks. You will not find many, because this bird rarely spends much time on the ground, preferring vantage points at the tops of trees or cliffs, where it also likes to build its huge nests.

Similar Species: The Turkey Vulture (*Cathartes aura*) has tracks of similar size. There are many smaller birds of prey with similar prints in smaller sizes.

Great Horned Owl

Strike
Width: to 3 ft (90 cm)
Size
22 in (55 cm)

GREAT HORNED OWL
Bubo virginianus

Often seen resting quietly in trees by day, this wide-ranging owl prefers to hunt at night. An accomplished hunter in snow, the owl strikes through the snow with its talons, leaving an untidy hole that may be surrounded by wing and tail-feather imprints. If it registers well, this 'strike' can be quite a sight. The feather imprints are made as the owl struggles to take off with possibly heavy prey. An ungraceful walker, it prefers to fly away from the scene.

You may stumble across a strike and guess that the owl's target could have been a vole (p. 92) scurrying around underneath the snow. Or you may be following the surface trail of an animal to find that it abruptly ends with this strike mark, where the animal has been seized.

Similar Species: If the prey left no approaching trail, the strike mark is likely an owl's, because owls hunt by sound. If there is a trail, the strike mark, usually with less rounded and more distinct feather imprints, could be by a hawk or a Common Raven (p. 122), both of which hunt by sight.

Common Raven

Print
Length: to 4 in (10 cm)
Straddle
4 in (10 cm)
Stride
Walking: to 6 in (15 cm)
Size
2 ft (60 cm)

COMMON RAVEN
Corvus corax

This legendary bird spends a lot of time strutting around on the ground—confident behavior that may hint at its intelligence.

Ravens leave a typical alternating track pattern. Prints show three thick toes pointing forward and one toe pointing backward. When a Raven is in need of greater speed, perhaps for takeoff, it leaves a trail of diagonally placed pairs of prints that are rather irregular.

Similar Species: Other corvids, such as the American Crow (p. 124), also spend a lot of time poking around on the ground. Their tracks are similar, but smaller, and their strides are correspondingly shorter.

American Crow

Print
Length: 2.5–3 in (6.5–7.5 cm)
Straddle
1.5–3 in (3.8–7.5 cm)
Stride
Walking: 4 in (10 cm)
Size
16 in (40 cm)

AMERICAN CROW
Corvus brachyrhyncos

The black silhouette of the American Crow can be a common sight in a variety of habitats. Like the Common Raven (p. 122), the American Crow will frequently come down to the ground and contentedly strut around. Its loud *caw* can be heard from quite a distance. Crows can be especially noisy when they are mobbing an owl or a hawk.

The American Crow typically leaves an alternating walking track pattern. Its prints show three sturdy toes pointing forward and one toe pointing backward. When a crow is in need of greater speed, perhaps for takeoff, it bounds along, leaving irregular pairs of diagonally placed prints with a longer stride between each pair.

Similar Species: Other corvids, such as jays, also spend a lot of time on the ground and make similar tracks. The much larger Common Raven leaves similar but larger tracks.

Dark-eyed Junco

Print
Length: to 1.5 in (3.8 cm)
Straddle
1–1.5 in (2.5–3.8 cm)
Stride
Hopping: 1.5–5 in (3.8–13 cm)
Size
5.5–6.5 (14–17 cm)

DARK-EYED JUNCO

Junco hyemalis

This common small bird typifies the many small hopping birds found in the region. Each foot has three forward-pointing toes and one longer toe at the rear. The best prints are left in snow, although in deep snow the toe detail is lost; the footprints may show some dragging between the hops.

A good place to study this type of prints is near a birdfeeder. Watch the birds scurry around as they pick up fallen seeds, then have a look at the prints left behind. For example, Dark-eyed Juncos are attracted to seeds that chickadees (*Poecile* spp.) scatter as they forage for sunflower seeds in the birdfeeder. Also look for tracks under coniferous trees, where juncos feed on fallen seeds in winter.

Similar Species: Toe size may help with identification — larger birds make larger prints — as can the season. Junco tracks in powdery snow could be mistaken for mouse (pp. 94–97) tracks, so follow the trail to see if the tracks disappear down a hole or into thin air.

Frogs

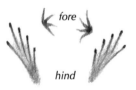

fore

hind

Straddle
to 3 in (7.5 cm)

hopping

FROGS

Bullfrog

The smallest frogs include the treefrogs. The Spring Peeper's (*Pseudacris crucifer*) 1.5-inch (3.8-cm) length and its preference for thick undergrowth and shrubs near the water make its tracks a rare sight. The larger gray treefrogs (*Hyla chrysoscelis* and *H. versicolor*) spend most of their time in trees, coming down to breed and sing at night. The Green Frog (*Rana clamitans*) and the Pickerel Frog (*R. palustris*) both favor slow-moving, shallow water and swampy areas, and both are widespread. The beautiful Southern Leopard Frog (*R. sphenocephala*), to 5 inches (13 cm) long, makes larger tracks, but it is only found in the southern Appalachians. Unusually large tracks are surely from the robust Bullfrog (*R. catesbeiana*). Growing to 8 inches (20 cm) in length, it is North America's largest frog.

A frog's hopping action results in its two small forefeet registering in front of its long-toed hind prints. Frog tracks vary greatly in size, depending on species and age. The best place to look for frog tracks is along the muddy fringes of waterbodies. Toads (p. 130) may also hop, but usually they walk.

Toads

hind   *fore*

Straddle
to 2.5 in (6.5 cm)

walking

TOADS

Woodhouse's
Toad

There are fewer toad species than frog species in the Appalachian region. Undoubtedly, the best place to look for toad tracks is, as with frog tracks, along the muddy fringes of waterbodies, but these tracks can occasionally be found in drier areas, for example, as unclear trails in dusty patches of soil.

The toad most likely to be encountered, and the most widespread, is the American Toad (*Bufo americanus*), which lives in many different moist habitats. Woodhouse's Toad (*B. woodhousei*) is found scattered throughout the region in temporary pools and ditches. Toads in this region can be up to 4.5 inches (11 cm) in length.

In general, toads walk and frogs (p. 128) hop, but toads are pretty capable hoppers, too, especially when being hassled by overly enthusiastic naturalists. Toads leave rather abstract prints as they walk. The heels of the hind feet do not register. On less firm surfaces, the toes often leave draglines.

Salamanders & Newts

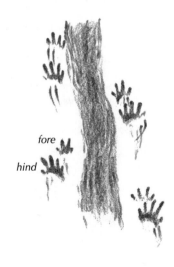

fore

hind

Straddle
to 3 in (7.5 cm)

walking

SALAMANDERS & NEWTS

Eastern Tiger Salamander

There are a wealth of salamanders and newts in the moist and wet areas of the Appalachians. Among the more abundant and widespread of these long, slender, lizard-like amphibians is the Eastern Newt (*Notophthalmus viridescens*), which can grow to 5.5 inches (14 cm) in length. After a fresh rain, Eastern Newts emerging from ponds may leave small trails in the mud.

The Spotted Salamander (*Ambystoma maculatum*), which can grow to 10 inches (25 cm) long, lives throughout the region in mixed forests and some coniferous forests. In moist ravines and canyons, a lucky observer might find evidence of the Zigzag Salamander (*Plethodon dorsalis*). The king of the salamander world is undoubtedly the magnificent Eastern Tiger Salamander (*A. tigrinum*), which can grow up to 13 inches (33 cm) long and comes in such a diversity of colors and patterns that it defies description. This heavy salamander leaves the best tracks, with a straddle of up to 4 inches (10 cm).

In general, a salamander fore print shows four toes, and the larger hind print shows five. However, print detail is often blurred by the animal's dragging belly or by the swinging of its thick tail across the tracks.

Lizards

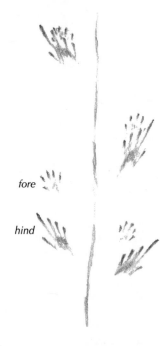

fore

hind

Straddle
to 3 in (7.5 cm)

walking

LIZARDS

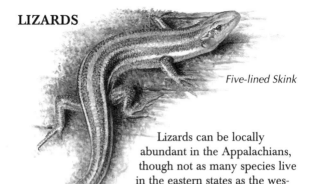

Five-lined Skink

Lizards can be locally abundant in the Appalachians, though not as many species live in the eastern states than the western states. Lizards leave tracks similar to those of the salamanders, but their toes are longer and more slender.

If you find lizard tracks, a likely candidate is the Five-lined Skink (*Eumeces fasciatus*), which can grow up to 8 inches (20 cm) long. It favors moist woodlands and is widespread throughout the Appalachians. Another skink that you might encounter is the Broadhead Skink (*E. laticeps*).

The Racerunner (*Cnemidophorus sexlineatus*) is also a likely suspect. This lizard, which can reach 11 inches (28 cm) in length, favors dry grasslands and well-drained woodlands. The Fence Lizard (*Sceloporus undulatus*), which grows to 8 inches (20 cm) in length, is common in open woodlands and grasslands.

These reptiles all move very quickly when the need arises, their feet barely touching the ground as they dart for cover. Consequently, their tracks can be hard to make out clearly.

Turtles

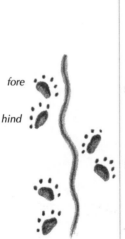

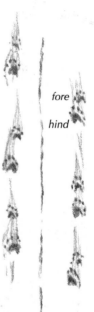

fore

hind

fore

hind

Straddle
4–10 in (10–25 cm)

Snapping Turtle walking

typical turtle walking

TURTLES

Eastern Box Turtle

Turtles, those ancient inhabitants of the water world, will happily slip into the murky depths to avoid detection. They do, however, come out from time to time to feed or to bask in the sunshine. Look for their distinctive tracks alongside ponds, rivers and moist areas. Some turtles, such as the huge Common Snapping Turtle (*Chelydra serpentina*), prefer to stay in the water and rarely come out.

One of the most widespread and prettiest turtles, often seen basking, is the Painted Turtle (*Chrysemys picta*); it can reach almost 10 inches (25 cm) in length. Slightly smaller is the well-named Common Musk Turtle (*Sternotherus odoratus*), which lives in shallow water and produces a foul odor. The Slider (*Trachemys scripta*), which reaches 12 inches (30 cm) in length, prefers slow-moving rivers. The Eastern Box Turtle (*Terrapene carolina*) is common in forested areas.

With its large shell and short legs, a turtle has a straddle that is about half its body length. Although longer-legged turtles can raise their shells off the ground, short-legged species may let them drag, as shown in their tracks. The tail may leave a straight dragline in the mud. On firmer surfaces, look for distinct claw marks.

Snakes

SNAKES

Common Garter Snake

Many snake species inhabit the Appalachians, but with much greater diversity in the warmer south. Because snakes are all long and slender, their tracks appear so similar that identification among the species is next to impossible. In fact, because a snake lacks feet and leaves a track that is just a gentle meander, it is very challenging even to establish in which direction a snake was moving.

The most widespread and frequently encountered snake is the harmless Common Garter Snake (*Thamnophis sirtalis*). Found throughout the region, often close to wet or moist areas, it can reach 4.3 feet (1.3 m) in length. Also widespread is the Redbelly Snake (*Storeria occipitomaculata*), which prefers hilly woodlands and can grow to 16 inches (40 cm) long. Inhabiting grassy meadows and fields along forest edges, the Rough Green Snake (*Opheodrys aestivus*) can grow to 3.8 feet (1.2 m) in length. The rattlesnake most likely to be encountered is the Timber Rattlesnake (*Crotalus horridus*), which can grow to 6.3 feet (1.9 m) long; it frequents a variety of habitats from marshlands to dry woodlands. Also common in a variety of habitats is the large and colorful Common Kingsnake (*Lampropeltis getula*).

TRACK PATTERNS & PRINTS

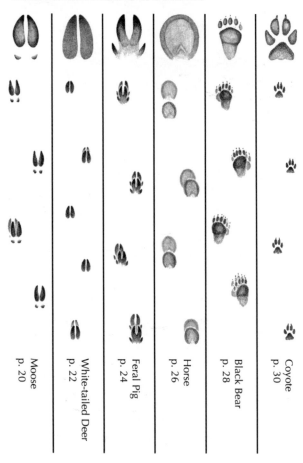

Moose
p. 20

White-tailed Deer
p. 22

Feral Pig
p. 24

Horse
p. 26

Black Bear
p. 28

Coyote
p. 30

Red Fox
p. 32

Gray Fox
p. 34

Bobcat
p. 36

Domestic Cat
p. 38

Raccoon
p. 40

Opossum
p. 42

TRACK PATTERNS & PRINTS

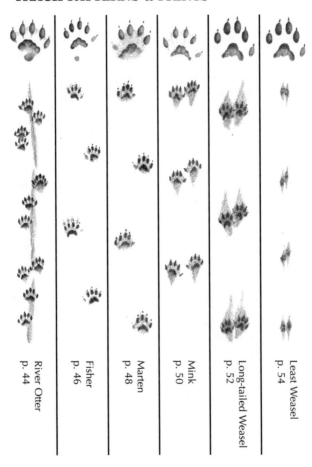

River Otter
p. 44

Fisher
p. 46

Marten
p. 48

Mink
p. 50

Long-tailed Weasel
p. 52

Least Weasel
p. 54

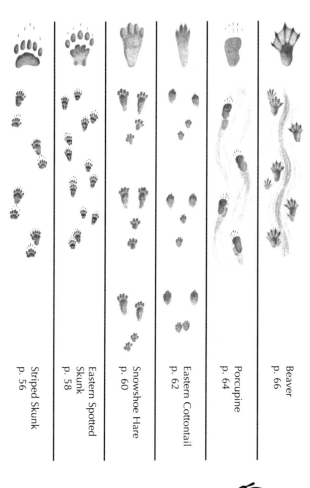

TRACK PATTERNS & PRINTS

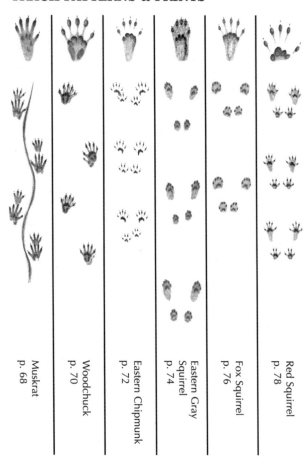

Muskrat
p. 68

Woodchuck
p. 70

Eastern Chipmunk
p. 72

Eastern Gray
Squirrel
p. 74

Fox Squirrel
p. 76

Red Squirrel
p. 78

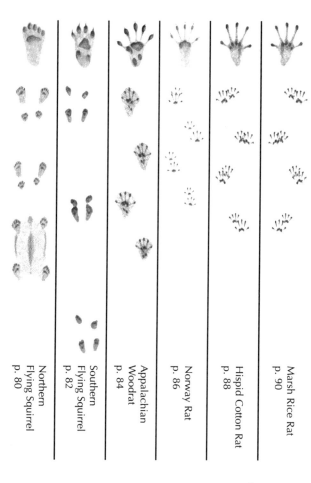

Northern
Flying Squirrel
p. 80

Southern
Flying Squirrel
p. 82

Appalachian
Woodrat
p. 84

Norway Rat
p. 86

Hispid Cotton Rat
p. 88

Marsh Rice Rat
p. 90

TRACK PATTERNS & PRINTS

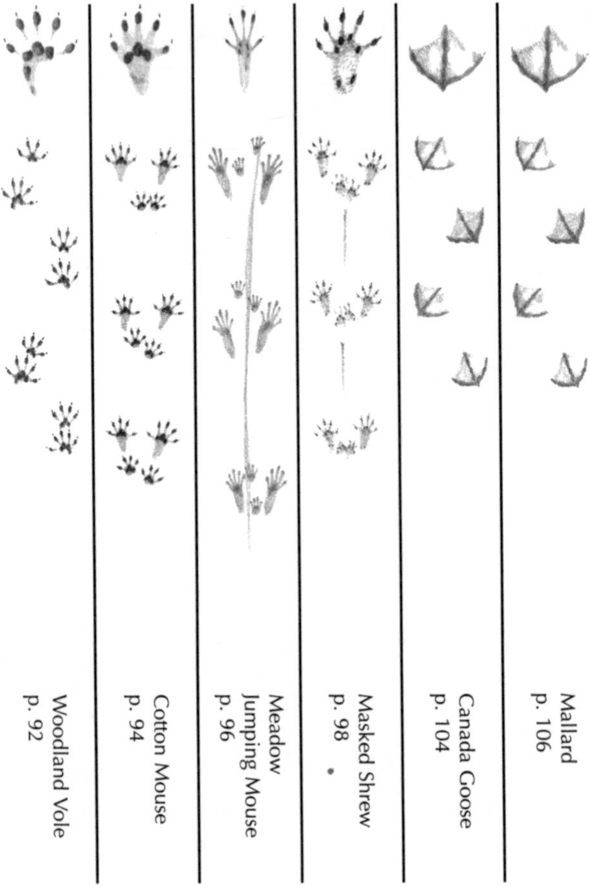

Woodland Vole
p. 92

Cotton Mouse
p. 94

Meadow
Jumping Mouse
p. 96

Masked Shrew
p. 98

Canada Goose
p. 104

Mallard
p. 106

146

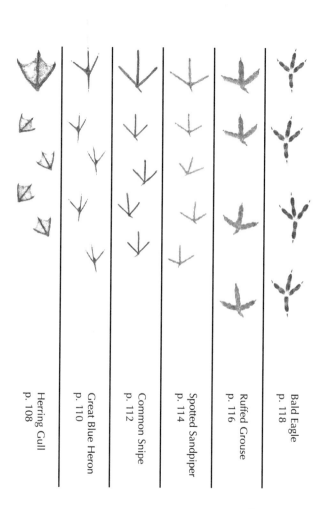

147

TRACK PATTERNS & PRINTS

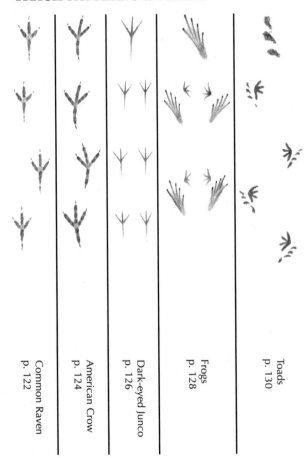

Common Raven
p. 122

American Crow
p. 124

Dark-eyed Junco
p. 126

Frogs
p. 128

Toads
p. 130

Salamanders
p. 132

Lizards
p. 134

Turtles
p. 136

Common Snapping
Turtle
p. 136

Snakes
p. 138

HOOFED PRINTS

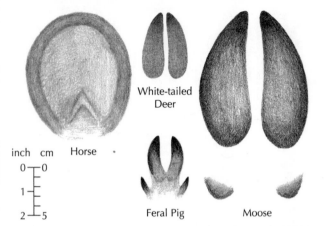

inch cm

Horse

White-tailed Deer

Feral Pig

Moose

HIND PRINTS

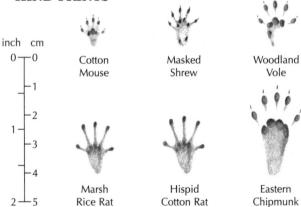

inch cm

Cotton Mouse

Masked Shrew

Woodland Vole

Marsh Rice Rat

Hispid Cotton Rat

Eastern Chipmunk

HIND PRINTS

Southern
Flying
Squirrel

Meadow
Jumping
Mouse

Norway
Rat

Appalachian
Woodrat

Northern
Flying
Squirrel

Red
Squirrel

Woodchuck

Muskrat

Fox
Squirrel

Eastern
Gray
Squirrel

Opossum

Raccoon

Porcupine

Eastern Cottontail

Snowshoe Hare

inch   cm

0 ─┬─ 0

1 ─┼─

2 ─┴─ 5

HIND PRINTS

inch cm
0 — 0
2
4 — 10

Beaver

Black Bear

FORE PRINTS

Least
Weasel

Long-tailed
Weasel

Mink

Domestic
Cat

Bobcat

Eastern
Spotted
Skunk

Striped
Skunk

Marten

Gray
Fox

Red
Fox

inch cm
0 — 0
1
2 — 5

River Otter

Fisher

Coyote

152

BIBLIOGRAPHY

Behler, J.L., and F.W. King. 1979. *Field Guide to North American Reptiles and Amphibians.* National Audubon Society. New York: Alfred A. Knopf.

Brown, R., J. Ferguson, M. Lawrence and D. Lees. 1987. *Tracks and Signs of the Birds of Britain and Europe: An Identification Guide.* London: Christopher Helm.

Burt, W.H. 1976. *A Field Guide to the Mammals.* Boston: Houghton Mifflin Company.

Farrand, J., Jr. 1995. *Familiar Animal Tracks of North America.* National Audubon Society Pocket Guide. New York: Alfred A. Knopf.

Forrest, L.R. 1988. *Field Guide to Tracking Animals in Snow.* Harrisburg: Stackpole Books.

Halfpenny, J. 1986. *A Field Guide to Mammal Tracking in North America.* Boulder: Johnson Publishing Company.

Headstrom, R. 1971. *Identifying Animal Tracks.* Toronto: General Publishing Company.

Murie, O.J. 1974. *A Field Guide to Animal Tracks.* The Peterson Field Guide Series. Boston: Houghton Mifflin Company.

Rezendes, P. 1992. *Tracking and the Art of Seeing: How to Read Animal Tracks and Signs.* Vermont: Camden House Publishing.

Stall, C. 1989. *Animal Tracks of the Rocky Mountains.* Seattle: The Mountaineers.

Stokes, D., and L. Stokes. 1986. *A Guide to Animal Tracking and Behaviour.* Toronto: Little, Brown and Company.

Wassink, J.L. 1993. *Mammals of the Central Rockies.* Missoula: Mountain Press Publishing Company.

Whitaker, J.O., Jr. 1996. *National Audubon Society Field Guide to North American Mammals.* New York: Alfred A. Knopf.

INDEX

Page numbers in **boldface** type refer to the primary (illustrated) treatments of animal species and their tracks.

155

ABOUT THE AUTHORS

Tamara Eder, equipped from the age of six with a canoe, a dip net and a note pad, grew up with a fascination for nature and the diversity of life. She has a degree in environmental conservation sciences and has photographed and written about the biodiversity in Bermuda, the Galapagos Islands, the Amazon Basin, China, Tibet, Vietnam, Thailand and Malaysia.

Ian Sheldon, an accomplished artist, naturalist and educator, has lived in South Africa, Singapore, Britain and Canada. Caught collecting caterpillars at the age of three, he has been exposed to the beauty and diversity of nature ever since. He was educated at Cambridge University and the University of Alberta. When he is not in the tropics working on conservation projects or immersing himself in our beautiful wilderness, he is sharing his love for nature. Ian enjoys communicating this passion through the visual arts and the written word.